素色美学空间

2018-2019 现代设计奖获奖作品集

艺力国际出版有限公司　编

岭南美术出版社

中国·广州

图书在版编目（ＣＩＰ）数据

素色美学空间：2018-2019现代设计奖获奖作品集 =
Modern Plain Aesthetics Space：Collection of
Awarded Works for Modern Design Award 2018-2019 /
艺力国际出版有限公司编. — 广州：岭南美术出版社，
2020.3

ISBN 978-7-5362-6979-8

Ⅰ．①素… Ⅱ．①艺… Ⅲ．①空间设计－作品集－世
世－现代 Ⅳ．①TU206

中国版本图书馆CIP数据核字(2020)第014371号

出 版 人：李健军
责任编辑：刘向上　　张柳瑜
责任技编：罗文轩
特约编辑：李爱红　　沈敏萍　　王琛
美术编辑：陈　婷
翻　译：莫婷丽　　彭小竹

素色美学空间：2018-2019现代设计奖获奖作品集 = Modern Plain Aesthetics
Space：Collection of Awarded Works for Modern Design Award 2018-2019
SUSE MEIXUE KONGJIAN：2018-2019 XIANDAI SHEJIJIANG HUOJIANG ZUOPINJI

出版、总发行：岭南美术出版社　　（网址：www.lnysw.net）
　　　　　　　　（广州市文德北路170号3楼　邮编：510045）
经　　　销：全国新华书店
印　　　刷：深圳市汇亿丰印刷科技有限公司
版　　　次：2020年3月第1版
　　　　　　　2020年3月第1次印刷
开　　　本：960 mm×1194 mm　1/16
印　　　张：22.5
字　　　数：51千字
ISBN 978-7-5362-6979-8

定　　　价：358.00元

Preface / 序言

"A new generation of designers is enlarging their world little by little, which can be seen from the works in this book. And because of their enthusiasm to design, modern design can be more creative and promising".

Design serves people. It has the power to arouse one's emotions and feelings. For me, space design is the aesthetics of handling relations among human, space and nature. And the creation of a harmonious relation depends on the nicely placement of the three.

Gradually, space design becomes the synthesis of life and aesthetics with the integration of originality, colors and cultural aesthetics. We can see the human landscape in design and touch the art of light and shadow. Design makes us regain power after a break in peacefulness by relieving our pain, cleaning our mind and harboring our soul. Space design is not only to make a space hold various forms and possibilities, but also to bring art and life together.

Zhang Qingping

"在本次的竞图作品中，我看到许多新一代设计者正一点一点向外扩展自己的世界，正因为有这群对设计热爱的人们，现代设计才能不断创新、不断进步"。

设计是为人们而造的，必须具备唤起人们情感，激发人们感受力的能力。我认为空间设计就是处理关系的美学，将人与人、人与空间、人与自然，放在彼此适当的距离，创造出最美好的关系。

原创、色彩、与人文美学的注入，使得空间设计成为生活与美学的综合体。我们在设计中遇见人文风景，感受光影交织的艺术。设计为我们的心灵带来庇护，具备疗愈力，滤净杂思，让人清明，安然宁静，能让人们在短暂的休息后，再次充满力量。空间设计不只让空间容纳多种形式与多种可能，同时也让艺术与生活交会。

张清平

Contents / 目录

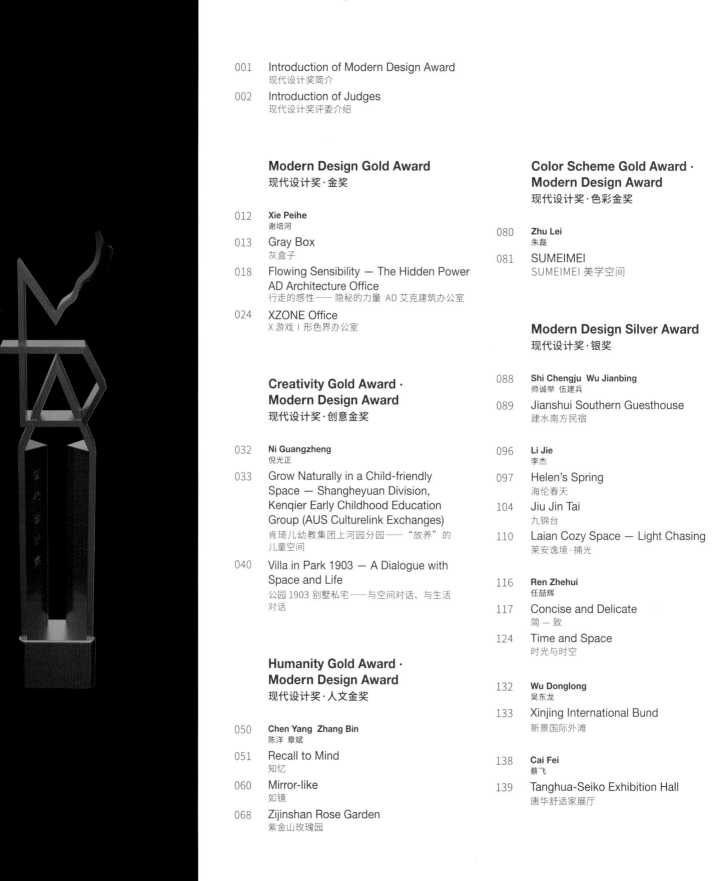

Introduction of
Modern Design Award
现代设计奖简介

International Youth Designers Association, joining hands with Overland Ceramics Co.,Ltd. and other groups from design industry, established the Modern Design Award to discover passionate potential new designers and provide them an international platform, which further can encourage young designers to study how design influences modern dwelling culture, and therefore, to reveal a modern design trend with both Chinese and western philosophy.

"Modern Design Award" covers a wide range of commercial, residential and public spaces with particular modern design style. And based on multiple dimensions like creativity, color, aesthetics, humanity and personality, its expert advisors from IYDA (International Youth Designers Association) will evaluate and select design works to be awarded the Modern Design Gold Award, Creativity Gold Award, Humanity Gold Award, Color Scheme Gold Award, Modern Design Silver Award, Design Aesthetics Award and Modern Design Excellence Award.

为推介培育热衷设计、具有潜质的新生代设计师，并为其提供国际化的成长环境和展示平台，国际青年设计协会联合设计业界，由欧文莱素色现代砖支持，共同发起举办"现代设计奖"。鼓励青年设计师群体研究设计对现代人居文明的影响，提炼并体现兼容中西哲学的现代设计思潮。

"现代设计奖"专注现代设计风格领域，集结国际青年设计师协会专家顾问团作为评委矩阵，覆盖商业空间、住宅空间、公共空间，从创意、色彩、美学、人文、个性等多维进行甄选，产生七大奖项：现代设计奖·金奖、现代设计奖·创意金奖、现代设计奖·人文金奖、现代设计奖·色彩金奖、现代设计奖·银奖、现代设计奖·美学奖、现代设计奖·优秀奖。

Introduction of Judges

现代设计奖评委介绍

Zhang Qingping 张清平

Founder and Chief Designer of Tienfun Interior Planning LTD.
Director-general of Taiwan National Association Interior Designers

天坊室内计划创始人 & 总设计师
台湾室内设计专技协会理事长

Modern design involves designers' attitudes toward the proper expression of details as well as sharp observation on beauty. It also combines classic symmetry and modern proportion, western gracefulness and eastern sumptuousness. Just as Zhang Qingping said when he talked about the true meaning of luxury, to find our way in a journey is happiness, similarly, to experience wonderfulness when we lose our way is also happiness but more like a luxury. He modestly clarified that he wasn't born to be a talented designer, and he always brings a measuring tape with him each time he travels abroad. As soon as he arrives at the hotel, he measures the room, then it is time to enjoy the feelings that the space brings to him. After continuous learning and accumulation, he finally returns all those wonderful feelings to the owners through the spaces he designed. He deeply knows that one cannot make himself a true master within a day, so he always keeps his original spirit and attitude to space. At last, his perseverance brings us not only visual enjoyment in luxurious spaces, but also tactual enjoyment of luxury of master style.

　　现代设计以一种极致表现细节的态度与对美感的敏锐，成功揉合古典对称、现代比例、西方优雅与东方奢华。张清平用一句贴近心灵的话语点醒奢华的真义"在旅行中找到方向是幸福，在迷途中发现精彩是奢华的幸福"，张清平谦虚地说自己并非天才型的设计师，因为总是随身携带著卷尺，每每出国旅行一抵达饭店第一件事就是丈量房间尺寸，然后再静静体验空间带来的感觉，透过不断地学习研究累积，将美好的感受回馈到屋主悉心交付的空间中。即便早已知道大师成就并非一蹴可及，张清平对空间从一而终的精神及态度，让我们看到不只是空间的奢华，更能从他的设计中感受到拥有奢华内涵的大师风范。

Li
Jin 李劲

Vice-director of Ceramics Art Committee of
Guangdong Artists Association

广东省美术家协会陶瓷艺术委员会
副主任

Executive Director of Ceramic Arts Committee

Secretary of Art Committee of Guangdong Ceramics Association

Dean and Postgraduate Supervisor of School of Traditional Chinese
Painting, Guangzhou Academy of Fine Arts (GAFA)

Associate Dean of Lingnan Culture and Art Research Institute,
Guangzhou University

Senior Industrial Artist

Executive Director of Guangdong Arts and Crafts Association

Researcher of the Council of Folk Artists Association Of Guangdong

Honorary President of Huizhou Arts and Crafts Guild

Guest Professor of Guangzhou Senior Light Industry Technical School

Member of the Presidium of Foshan Artists Association

Curator of Guangzhou Strait Art Museum

Master of Guangdong Ceramic Arts

Member of China Artists Association

中国陶瓷工业协会陶瓷艺术委员会 常务理事
广东省陶瓷协会艺术委员会秘书长
广州美术学院中国画学院院长
硕士研究生导师
广州大学岭南文化艺术研究院副院长
广东省工艺美术协会常务理事
正高级工艺美术师
广东省民间文艺家协会理事研究员
惠州市工艺美术行业协会名誉会长
广州市轻工高级技工学校客座教授
佛山市美术家协会主席团成员
广州海峡美术馆馆长
广东省陶瓷艺术大师
中国美术家协会会员

Simon
Yu　　俞锦文

Design Director and Senior Associate of
Zaha Hadid Architects Hong Kong Office

扎哈·哈迪德建筑事务所香港办事处
设计总监 & 高级合伙人

Simon Joined Zaha Hadid Architects in 1995. He worked for it again in 2004 with rich international working experience some time later, and since then, he has been the Director of Zaha Hadid Architects Hong Kong Office.

　　Simon 于 1995 年首次加入扎哈·哈迪德建筑事务所，在累计一段时间的国际工作经验之后，于 2004 年再次加入，自此担任扎哈·哈迪德建筑事务所中国香港办事处的负责人。

Chen Ya'nan 陈雅男

Editor-in-Chief of Modern Decoration and Senior Media Personnel

《现代装饰》杂志主编、
资深媒体人

As the Editor-in-Chief of Modern Decoration and a senior media personnel as well as a judge of "International Design Media Award", she is good at capturing the trend of design industry and has devoted to the observation and research of design area and designers, the discovery of potential designers, and the development of design industry.

　　《现代装饰》杂志主编，资深媒体人，担任"国际设计传媒奖"媒体评委。多年来，一直专注于对设计领域的观察与研究，擅于捕捉设计行业发展动态，致力于以媒体人的敏锐观察与思辨能力大力挖掘设计新锐，助推设计向前发展。

Zhuang Chenghao 庄承浩

National Chief Editor of Sina Furniture Channel and
Chief Editor of Shenzhen Sina Home

新浪家具频道全国主编
兼深圳新浪家居主编

Zhuang Chenghao is the senior media personnel, exhibition planner, National Chief Editor of Sina Furniture Channel, Chief Editor of Shenzhen Sina Home, and guest contributor of many media platforms like Hong Kong Furniture (magazine), Furniture Today (America) and Furniture Micro-News (Wechat Platform). His critique *Witnessed the Unusual Phenomenon of Interior Design Cycle in a Decade* criticized that many designers tend to disparage each other, resort to publicity stunt and pursue fame and fortune, which went viral on WeChat moments. He was also the main planner of the H5 Project "Che Jianxin Invites You to Group Chat" which become a typical marketing event of home design and the new Chinese brand activity "American Ways of Living in 2020", which gained close attention in business circle including Dong Mingzhu (Chairwoman of the board of Gree).

　　庄承浩，资深媒体人、策展人，新浪家具频道全国主编兼新浪家居深圳主编，《香港家私》杂志、美国《今日家具》中文版、家具微新闻等媒体平台特约撰稿人。

　　撰写的评论文章《十年目睹之室内设计圈怪现状》，抨击设计圈文人相轻，作秀成性，求名求利等现象刷爆了设计人的朋友圈；主导策划的"车建新邀请你加入群聊"H5成为家居圈现象级的事件营销；策划的以"2020年美国人的生活"为切入点的新国货品牌活动，引起格力董明珠在内企业界关注。

Liang Xueqing 梁雪青

General Manager of Brand Management Center,
Guangdong Overland Ceramics Co., Ltd.

广东欧文莱陶瓷有限公司品牌
管理中心总经理

Graduated from Guangzhou Academy of Fine Arts, Liang Xueqing once served as a planner of a famous Guangzhou 4A advertising company. In 2008 he joined Overland — one of the famous Chinese high-end ceramic tile export brands. With his professional design and planning perspective, he contributes to the building of the core brand value of Overland and the brand positioning of a creative modern solid-color ceramic tile.

In 2017, he creatively used the gray theory in his symbolic product "Overland Gray" which shocked the whole ceramic industry, then it become a grayscale reference standard of Chinese high-end ceramic tile, which also benefits the technological progress in this industry.

In 2018, a series of events were organized to promote the interaction between Overland and design industry, such as the 2018 International Designers Summit — aiming to establish the Modern Design Award to provide an international environment and platform for a new generation of designers.

　　毕业于广州美术学院，曾担任广州知名 4A 广告公司策划师。自 2008 年加入欧文莱，将专业的设计、策划视角，融入以产品见长、被誉为"中国高端瓷砖出口前列品牌之一"的欧文莱陶瓷。建构了欧文莱品牌聚焦的核心内容，塑造了创新型的素色现代砖品牌定位。

　　2017 年，打造的超级符号产品"欧文莱灰"风靡整个陶瓷行业，创新灰度理论，成为中国高级灰瓷砖的灰度参照标准，推动了行业的技术发展。

　　2018 年，主导推动了欧文莱品牌与设计界的一系列互动交流，2018 国际千人设计师峰会揭开"现代力量"序幕，重力打造现代设计奖，积极为新生代设计师提供国际化的成长环境和展示平台。

Lu Jican 卢积灿

Executive President of International Youth Designers Association
Co-founder of ACS Creative Space Design Agency
Publisher of Artpower International Publish Co., Ltd.

国际青年设计师协会执行会长、
ACS 创意空间设计机构联合创始人
艺力国际出版有限公司出版人

Specialized in publishing books on architecture, interior design and art for over ten years, he has unique perspective and art taste in space design, which drives him to bring out new trend and broader vision for art and design industry.

　　十余年专注于建筑室内艺术领域图书出版，对空间设计拥有自己独到的审美主张与艺术品位，始终坚持开创艺术设计领域的新思潮，致力于开拓设计行业更为广阔的视野。

Lu *Jican* 卢积灿

Executive President of International Youth Designers Association
Co-founder of ACS Creative Space Design Agency
Publisher of Artpower International Publish Co., Ltd.

国际青年设计师协会执行会长、
ACS 创意空间设计机构联合创始人
艺力国际出版有限公司出版人

Specialized in publishing books on architecture, interior design and art for over ten years, he has unique perspective and art taste in space design, which drives him to bring out new trend and broader vision for art and design industry.

　　十余年专注于建筑室内艺术领域图书出版，对空间设计拥有自己独到的审美主张与艺术品位，始终坚持开创艺术设计领域的新思潮，致力于开拓设计行业更为广阔的视野。

Modern Design Gold Award
现代设计奖·金奖

Xie
Peihe

谢培河

现代设计奖·金奖

Chief Designer of Shantou AD Architecture Design Co., Ltd.
2018–2019 the only winner of Modern Design Gold Award
2019 Red Dot Design Award (Germany)
2019 Frame Awards (Netherland)
2018 Best of Year Awards (America)
2018 Architecture Master Prize (AAP) (America)
2018 FX International Interior Design Awards (Britain)
2018 A`Design Award (Italy)

汕头市艾克建筑设计有限公司总设计师
曾获荣誉：
2018-2019 现代设计奖金奖唯一得主
德国 2019 红点设计大奖
荷兰 2019 Frame Awards 室内设计大奖
美国 2018 Best of Year Awards 大奖
美国 2018 建筑大师奖
英国 2018 FX 国际室内设计大奖
意大利 2018 A ` 设计大奖

Gray Box
灰盒子

设计机构：AD ARCHITECTURE ｜艾克建筑设计
主案设计师：谢培河
地点：广东汕头
面积：250 m²
主要材料：科技木饰面、瓷砖、灰玻璃、石材、亚光白漆
摄影：欧阳云

Hidden in a distant costal village in Shantou, Gray Box is an addition project to a rural self-built house. It is a high-level gray minimalist residence. The whole space is based on the most low-key black, white and gray, creating a texture space. Gray is highlighted to balance the strong contrast between black and white. Partial use of warm earth yellow penetrates temperature in Gray Box.

Its designer emphasizes the function of the space without intricate decoration to meet the owner's taste of life and health needs. Thoughtful choice of colors and materials in this space make it a simple, natural and comfortable home.

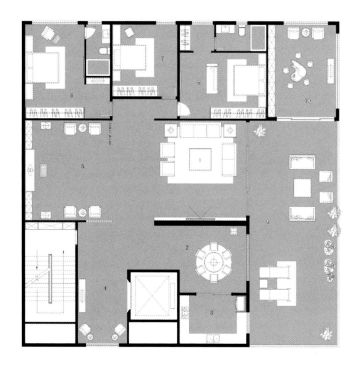

平面布置图　　　Plan

1. 客厅　　　　　1. living room
2. 餐厅　　　　　2. restaurant
3. 厨房　　　　　3. kitchen
4. 电梯厅　　　　4. elevator hall
5. 佛室　　　　　5. the temple
6. 主卧房　　　　6. master bedroom
7. 儿童房　　　　7. children's room
8. 长辈房　　　　8. elders room
9. 观景阳台　　　9. viewing balcony
10. 茶室　　　　10. tea room

提及农村自建屋，在很多人的固有印象里，都会出现一个即使花大价钱装修也依旧土里土气、把大把空间都浪费了的、几乎没什么设计感的独栋多层方格子。

灰盒子隐于远离闹市的汕头临海乡村，一个农村自建屋的加建项目。我们要发出另外一种声音，以黑白灰的极简与纯粹，为自建屋添上一道遗世独立的风景。业主年轻时尚，设计结合业主生活情节，打造高级灰极简住宅。整体以最为低调的黑白灰三色为主，营造质感空间。设计重笔在"灰"，含而不露，平衡着黑白界限分明的强烈对比。局部使用暖色系的大地黄，在灰盒子中渗透着温度。

极简，是一种情结，摒弃过多繁复的设计，反对刻板的形式设计态度，勿堆砌，给生活做减法，追求更适合人居住的生活环境。

极简强调空间的功能性，反对多余装饰。对色彩、材料的质感高要求，满足业主对生活品位和身体健康的需求，摆脱繁琐，追求简单，让整个家清新自然，干净舒适。

在这里，可以享受身心的放松，不用再耗费太多精力，只要在某处安静的空间找到独处的时光，放松自己的身心，感受静下来的感觉。

Flowing Sensibility — The Hidden Power
AD Architecture Office

行走的感性——隐秘的力量 AD 艾克建筑办公室

设计机构：AD ARCHITECTURE ｜艾克建筑设计
主案设计师：谢培河
地点：广东汕头
面积：850 m²
主要材料：金属钢材、亚光黑漆、水泥地板、科技木饰面、透明玻璃
摄影：欧阳云

The office of AD ARCHITECTURE is located in a creative park renovated from an old factory in Shantou.

In order to pursue the sense of scale brought by large space, we abandoned excessive partitions and adornments, and naturally extended the fusion of original space and new force through the composition of materials. The black paint finishes, cement floors and dark gray steel plates strengthened the consciousness of the original space's atmosphere. Without adopting unnecessary process and decoration, the additional built part is applied with industrialized steel and iron materials which corrode naturally as time passes

by. It seems that they are showing the essence of strength and beauty. The integrity and uniqueness of each piece of material are respected, and the structure is as simple and direct as possible.

Apart from preserving the original structure and satisfying the functional needs in the office, we also added some bold spaces where people can think quietly. Taking the public working area as the core, we created two functional areas to inject new vitality. The small attic formed by space division enhances the sense of form, and the originally tall space becomes very interesting through the intersection of different volumes. All volumes are seemingly connected

without interfering each other, and each space is provided with necessary functions.

The sun is power. A patio and French window are adopted to express the relation between human and nature while the original structure of building is retained. This simple window draws people closer to the exterior space. The patio is also another form of connection to the outside world. It not only ensures natural light but also enriches experience feelings in such a perfect space.

Lighting is also a major part of this design. We and tried to create light without revealing the lamps. Lighting strips were mostly used in the front desk and office area. The light and shadow in the empty ceiling area refract on the model planes again, which not only avoids deliberately decoration, but also creates a simple atmosphere and good lighting layout.

背景概括 | Background Summary

无论白天或黑夜，空间带来的体验感充满无尽的可能性，我们将它视为内心所追寻感性的力量，不遗不弃。因此这一次，一直坚持摒弃过多繁复设计理念的 AD 艾克建筑设计，为了追寻这份感性，将自己的办公空间打造成了设计灵魂的栖息地。

AD 艾克建筑办公空间坐落于汕头市，一座由老厂房改造的创意园里。这些老房子见证了这座城市的起落，随着城市迅猛的发展扩张，它们逐渐被遗忘或等待重新拆建改造。目睹着一幕幕的发生，我们仿佛能听到它们无力、落寞的诉说，不由让我们记起第一眼看到这空间时它所呈现出的原始生机和独特气质。难道它曾经带给我们有关于历史的诉说就这样结束了吗？当所有感性涌上心头，我们开始思考，并决定通过改造，让其重新呈现出属于它的使命。感知空间所赋予我们的灵感，并非过多去改变它，而是聆听它的诉说。在这一次设计中，尊重空间原有的意识形态，再注入新的活力，这正是我们需要去做好的事。

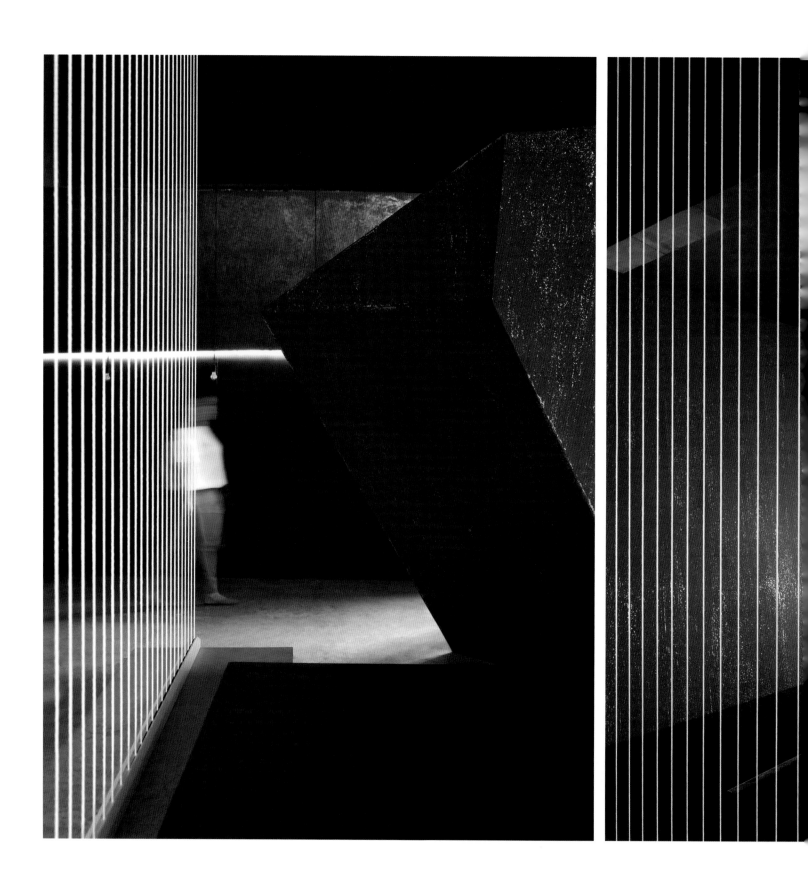

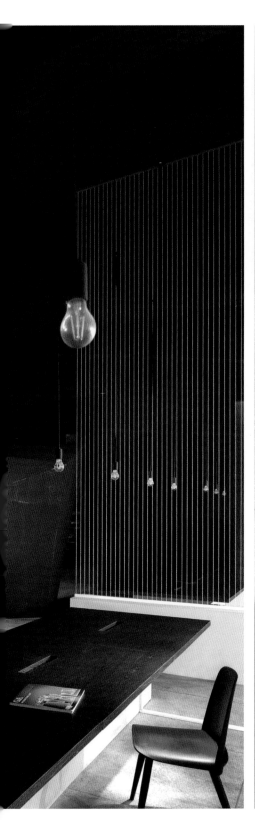

We selected simple materials as a respond to the space, so that the decorations are connected, and the individual character of the space is reflected. Great attention is paid to the attic, and at one side of the attic, a large person statue is looking at the door to welcome visitors to explore the undiscovered mysteries in this space. Walking closely, you can see the ball

which is made of iron wires is expressing art directly. The wires are interwoven with each other, just like designers' infinite thoughts. What we strove to do is to combine the artistic language to create a perceptual and powerful working space.

感知 | 原始空间的力量

在设计上，为了追求大空间所带来的尺度感，我们摒弃了过多的隔断和装饰，通过材质的构成，自然地延伸了原始空间与新力量的融合。黑漆饰面、水泥地板、深灰钢板强化了原始空间的意识氛围。用代表工业化的钢材、铁作为加建整体造型的材料，省略了无谓的加工和装饰，随着时间自然地锈蚀，就像试图展现力量与美的本质，尊重了每块材料的完整性和独特性。构造尽量简单而直接，无一不在诉说这空间的隐秘力量，强化凝聚了它原本的气质。

功能 | 增强形式感

在基本保留原始结构和满足办公功能使用的同时，我们在空间中增添了大胆的放空，大体是以公共办公区为核心，再分出两个戏码，注入了新的活力。空间分割出的一个小阁楼，增强了形式感，并通过体块穿插让原本高挑的空间变得具有趣味性，每个体块看似连贯，却又不相互制约；同时，让每个空间都具备其该有的功能形式。整体区域开敞舒适，关注点在于重生的形式感，我们希望这种形式感能延续到设计师工作的状态中，开放心态、敢于走出每一步；同时也是我们希望呈现给每个来访者的空间触感。

自然 | 活力的延伸

阳光即是力量。建筑本身的原构造被保留下来，我们用天井和简洁的大落地窗的形式去表达人与自然的关系，阳光从半透明的天井上洒向室内，在锈蚀的造型上倒映着阳光斑驳的影子，是活力的注入。简洁的大落地窗无意间成为了空间对外近距离的诉说，身处其中有着惬意的感受。随着折射的光影，我们能俯仰自然所带来的力量；设计师们在这里追逐梦想，唤醒对设计的坚持，也使想象力获得了自由。在繁琐的思绪里获得一丝平静，空间不再以单纯的形式主义而存在。在保留天井采光的设计上，我们旨在创造接近完美的空间，通过天井与外界形成的对话，丰富了空间的感受，也提供了更多不一样的体验感。

灯光 | 少即是多

灯光照明也是本次设计的重头戏码，坚持少即是多的原则，尽可能做到见光不见灯的状态。前台和办公区域的灯光照明，也多是利用灯带的方式去表达。挑空区光影和造型平面形成了第二次折射，没有做作的装饰氛围，既简约又能把握空间的光感尺度。

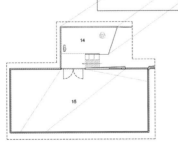

软装 | 艺术气质

　　软装作为装饰出现在空间设计中，是设计的符号，也是设计文化与生活的对接。我们在挑选软装上以材质简洁为基调，呼应空间，使得软装与空间有连续性，同时也体现了空间个性。小阁楼是本次装饰的着重点，不大的空间里每天都有很多的内心戏在上演。阁楼一侧偌大的人物雕像，望着门外，如同暗示着来访者关于这个办公空间还有未完的探险，走近，灯光下的铁丝球又在赤裸裸地表述了对艺术的诉说，一根根铁丝交织着，就像设计师的无限思绪。结合艺术的语言去塑造一个感性又有力量的办公间，是这一次我们所致力去做的。

XZONE Office

X 游戏 I 形色界办公室

设计机构：AD ARCHITECTURE I 艾克建筑设计
主案设计师：谢培河
项目团队：艾克创意
地点：广东汕头
面积：280 m²
主要材料：地坪漆、灰色乳胶漆（立邦）、混凝土板
摄影：欧阳云

With the contrast of square and round elements, the ingenuous utilization of materials, colors and formations, as well as application of abstract concept, the designer created a space full of unknowns.

A square box constitutes the main framework of the workspace. In the conversation area, the designer integrated semi-circle structures and a ball hung from the ceiling into the single spatial block, resulting in intersection of different boundaries. Thus it is a unique space with no rules to follow.

The designer adopted a restraint color scheme, with white and light gray as major hues, full of tension. Striking red and pink colors are dotted in the space, In the conference area, the red sofa and pink panel form contrast with the spatial overtone set by white color.

Materials of different textures break the consistency in the space and endow it with various features. The steel structure of the ceiling was retained, which becomes a highlight of the spatial design. The coordinated combination of rough and delicate textures generates contrast in the space, and creates a free atmosphere in it.

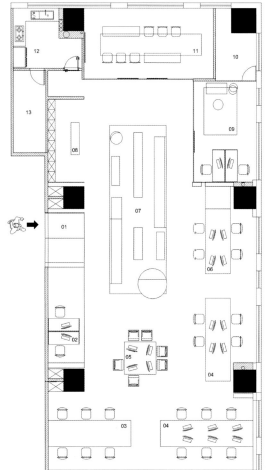

在空间整体冷静色彩的提炼中，具有无限张力的留白，这是设计师的克制处理。跳脱的红与粉，就像是这未知空间中躁动的原点。白本身处于色界的极端，在这界线分明的空间中，会议区成了克制的纯与空，红色沙发与粉色板块的激进表达，完全抛开了界的比例，对比鲜明。从色界极端入手，除了大胆的尝试，克制与躁动的较量，还交织着与情绪相辅的空间感受，引发人的行为特征，促进情感交流、思想碰撞。

未知，独立于"界"之外，是一种虚幻意象的根源，而虚幻意象促使我们去探索不被限定的界。在这意志力和注意力逐渐被弱化，感观愉悦不断被追捧的今天，我们似乎在用陌生、审视的目光注视着这座城市的中心，常常染上能够扼杀兴趣和消解愿意接受的"抑郁"。这个办公空间就犹如城市中的一座游乐场，每天都在营业。自由、模糊、突破、往外延伸，正示意着每场"未知"游戏的开始。

未知的永远在躁动，无拘于界限。

作品概述

X，是"未知"，是静是躁，是同是异，是不被界定。世界看似平静，却暗藏许多未知的可能，既和谐又矛盾，当我们放任自由，回归真我，不克制躁动的勇气，就能越过心中的界。矛盾、形式、构成、极致一同涌上，就如当下的我们，既简单又复杂。

设计理念

过滤心中的顾忌，创造躁动的"界"，跨越现实与理想，突破惯性思维，沉淀潜意识的力量，AD 艾克为"形色界"办公室探索出另一个未知的"界"。设计师在空间中强调方与圆的对比和冲突、极致的材料与形态的构成、克制的色彩与抽象的理念，创造性地表达出一个未知游戏。

形态不一 | 冲突与和谐

空间以方盒子作为主要框架，交谈区的半圆与单一体块共存，开放的形态空间延伸到顶上的圆球，形成不同边界的交汇。无规律可循，只有更多未知的组合。若有趣不一的元素相互碰撞，即使在不优质的空间里，也能赋有独一无二创造性。

材质对立 | 感受与灵感

通过材质的关联，天花与地面的分割打破了空间的一致性。天花保留了原建筑的钢筋结构，看似孤立的元素却无心成了最躁动的存在。通过原始的粗犷与细腻的形态构成，整个空间看上去都是自由的表达。既对立而又协调。

Creativity Gold Award ·
Modern Design Award
现代设计奖·创意金奖

Ni
Guangzheng

倪光正

现代设计奖 · 创意金奖

2018 Asia Pacific Interior Design Awards for Elite

曾获荣誉:
2018 亚太室内设计精英邀请赛大奖

Grow Naturally in a Child-friendly Space — Shangheyuan Division, Kenqier Early Childhood Education Group (AUS Culturelink Exchanges)

肯琦儿幼教集团上河园分园——"放养"的儿童空间

设计机构：LightingDesign 光正设计有限公司
主案设计师：倪光正

This project requires the designer to transform a commercial space into a children education space. The original space faces with many problems, such as partition, safety, lighting, ventilation and limitation of outdoor activity space. However, the terrace in the middle connects the three separate buildings to enhance the interflow among activity rooms and enlarge outdoor exercising space.

In a 1,800 m² indoor space, a widen white stair connects the first floor to the third floor. It is flat and safe for children to chase, play and jump after class. Each step is covered by gentle wooden boards to allow kids to sit here without feeling too cold.

There is no door for each classroom, tall walls among classrooms and the shared restrooms, which is a bold renovation and unexpectedly helps children to learn to follow rules and be polite. But in the middle of the classroom, the designer chooses low cabinets as partition wall to arouse children's curiosity,

A newly-built terrace becomes an open space for children to exercise and make friends. Viewed from above, you can see classrooms through the glass louver of the terrace. Children's napping zone is separated by gradient frosted glass panels which creates a tranquil and fairytale atmosphere. What's more, the space has not decorated excessive wiring tubes, plumbing, exposed cement beams and columns.

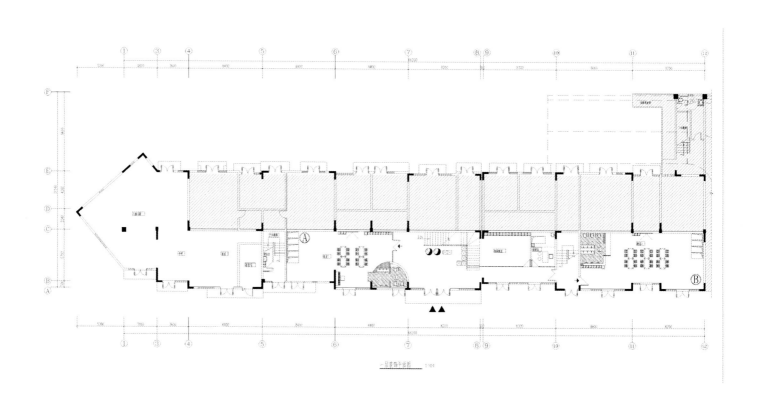

一层装饰平面图   1:100

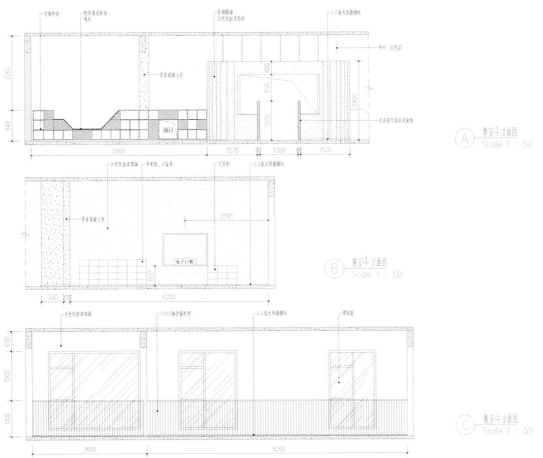

教室4 立面图
Ⓐ Scale 1 : 50

教室4 立面图
Ⓑ Scale 1 : 50

教室4 立面图
Ⓒ Scale 1 : 50

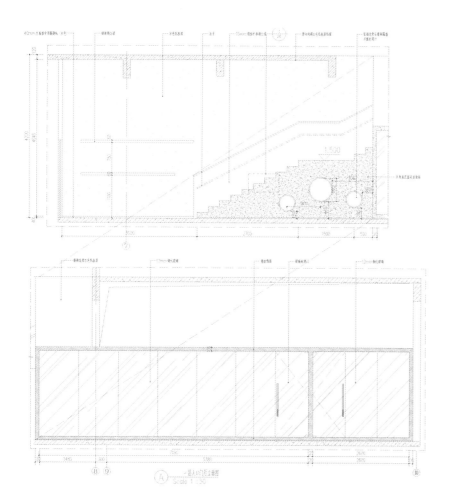

儿童教育空间的设计应该着眼于理解孩子的思维对于空间特征的反馈以及空间对于学习的提高作用上。环境和空间是基于对光线、声学、空气质量以及对于自然的观点这些能够深深地影响孩子的认知进程因素的理解之上，并赋予到空间中。

——约翰·保罗·艾伯哈德

成年人对于空间的认知主要是一种短暂的精神状态，而对于孩子来说，这种认知会对孩子意识的构建起到根本作用。表面单一的形态和色彩视觉美学已经远远不能满足当下幼儿教育的深度要求，从空间中获取知识，在空间中置入一些情景化的物品、色彩、图像，可以帮助小朋友构建自己的空间记忆，促进孩子智力上、感知上和行动上的激励，从而促使孩子强大自信心的建立和认知技能的发展。

将商业建筑空间改造成儿童教育空间，在原本固有的空间模数、安全因素、采光及通风、充裕的户外活动场地等方面都面临着很多短缺。利用中间的夹缝露台，将三幢独立的建筑进行搭建和连接，很好地促进了每间活动室交流通道的互连，扩大了户外活动场地，同时也增加了活动室的面积。

在改造完成的 1,800 m² 的室内空间中，孩子们的交流对象不再只是电视、手机和电脑，在这里可以完成在家里爸爸妈妈不让尝试的"危险"活动。连接一层到三层的是一部加宽加缓的白色楼梯，课余时间楼梯也成为小朋友追逐、嬉戏、跳跃的场所，柔和的木质踏步可以让小朋友尽情就地而坐，并且不会产生冰冷的不适感，也许这正是孩子与孩子之间无拘无束最好的交流方式。

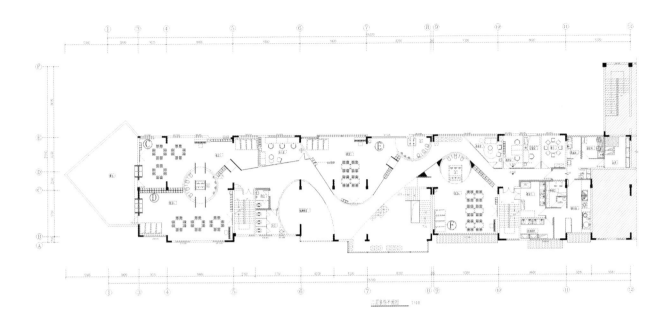

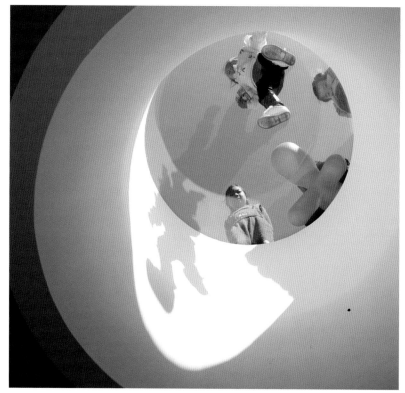

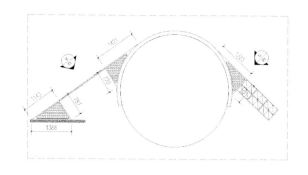

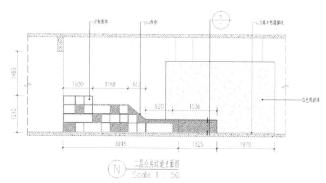

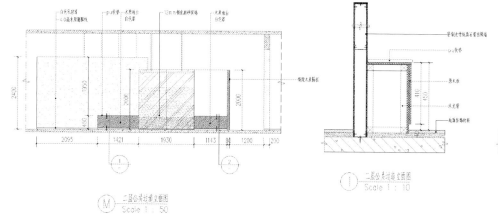

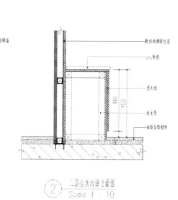

教室没有门，也没有班级与班级之间高高的隔墙，以及班级之间共用的卫生间，这是一次大胆的改造尝试，结果却让我们出乎意料，非但不乱，反而小朋友学会了遵守秩序和谦让。为了满足孩子的好奇心，教室中间的隔断矮柜便成为了他们的保护伞，课间偷偷越过矮柜一探隔壁教室又偷偷地爬回来。每个小朋友都可以用自己的方式开心地进入没有设门的教室，有攀爬进去的、越过矮柜跳进去的，就连平时性格内向、胆子很小的小朋友也跟着用这种不平常的方式来释放自己，很快地融入到集体中去。

搭建后的连接露台变成了小朋友对话和活动的户外场地，透过平台的玻璃天窗从上往下俯瞰教室，从下往上仰望天空的颜色、观察光线的变化是孩子们最开心的事儿。若隐若现的渐变磨砂玻璃隔断如同梦境一般，为小朋友的午休睡眠区增加了几分童话般的静谧。没有过多修饰的电线管、水管、暴露的水泥梁和柱，也是展现给每个孩子最真实、最自然的装点方式。

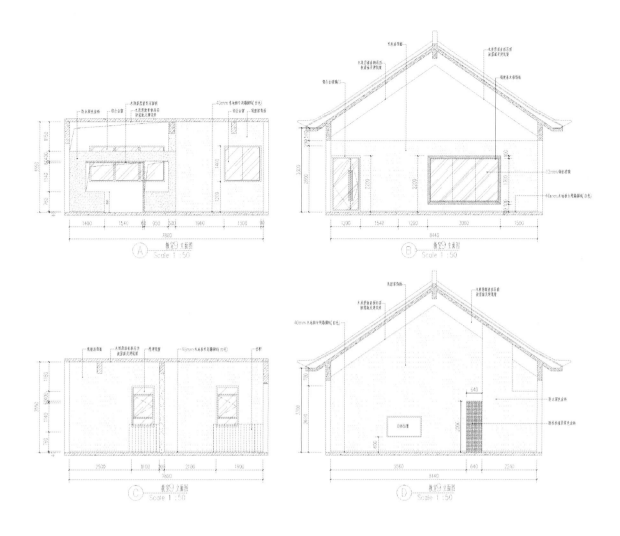

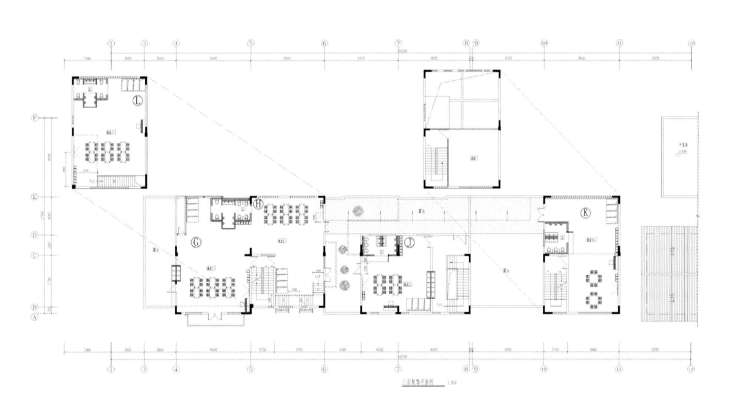

Villa in Park 1903 — A Dialogue with Space and Life

公园 1903 别墅私宅——与空间对话、与生活对话

设计机构：LightingDesign 光正设计有限公司
主案设计师：倪光正

This case covers an area of 450 square meters with two floors below and three floors above. To satisfy each family member's living requirement, the original functional areas are divided and rearranged to make full use of the landscape and daylight. Specific and proper storage of the life necessary is a good solution for the inconvenience caused by the vertical function distribution, and facilitates the organization and storage of goods. Easily found natural materials like rubbles and logs are well processed and combined to better fit the space, and then organically integrated with decorative and complementary artistic furnishings to create a natural, earthy and luxury atmosphere.

Passing through the parlor on the first floor, one can reach the backyard. And the renovated French window gives the best view of the garden scenery so that the static and boundless pool and aquatic plants are all visible and mounted at the eye level of people in the parlor, which also harmonizes the indoor environment with the green garden. Therefore, the boundary between indoor and outdoor spaces is blurred to create a natural, relaxed and quiet atmosphere. A renovated looming staircase in the middle not only guarantees the transparency and independency among the parlor, the children playing space and the living room, but also divides the space into three functional areas orderly and evenly.

A solid wood bar counter in the basement is extended as long as possible, which seems to be one step of the stairs. It also generates interesting functional visual aesthetics to harmonize the relation between space and human behavior.

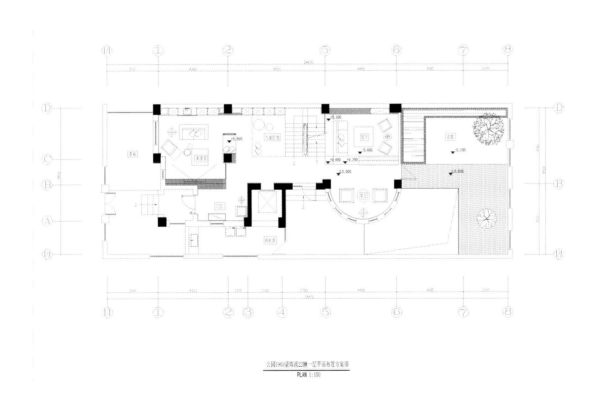

公園1903豪塔涅23棟 一层平面布置方案图
PLAN 1:100

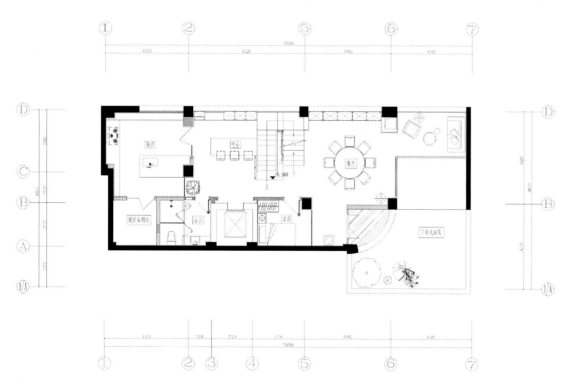

公园1903爱塔溪22栋头一层平面布置方案图
PLAN 1:100

在不以盲目追求风格、时尚和潮流为目的的前提下，将现代设计主义与传统的设计文化相结合，从空间中剥离多余的造型和色彩，把对传统文化的尊重和自然材料的欣赏转换为有温度、有文化、有内涵和有人情味的生活空间，于是便与主人产生了对话。

人的行为及感知与空间、物体、材质、光线、色彩的对话和互融是本次空间设计的思考重点。在这里，材质的自然性和细腻感如同人的肌肤，用温度和触感来激发人性化的空间，巧妙地成为人和空间的介质。

　　本案建筑面积为 450 m², 地下两层、地上三层, 为满足业主家庭结构成员的日常起居使用需求, 把原建筑功能位置进行拆分和打乱, 结合花园景观、采光等有利因素进行合理的二次功能布置。根据使用功能分区域和分类收纳生活起居用品, 很好地解决了垂直功能分布带来的使用非便捷性, 方便生活用品随时使用随时收纳归位。采用身边随处可见的朴素自然属性的材料 (天然碎石、自然木), 经过很好的加工和组合, 成为空间中的亮点, 自然朴实的奢华来源于具有艺术气质的点缀性、互补性陈设品和自然材料的有机结合。

　　一层会客厅是进出后花园的通过区域, 为了创造出花园景观和室内环境的互融氛围, 充分利用会客厅改造后的大落地窗, 保证花园里设置的静态无边水池和水生植物与会客厅里的使用者视线高度相同, 最大维度获取花园视野景观, 进而模糊了室内空间和室外景观的界限, 营造出一种自然、放松、安静的气氛。中间改造完成的若隐若现的楼梯, 在保证起居室、儿童活动区、会客厅的通透性和独立性的同时, 有序均匀地分隔了三个功能区。

　　最大限度地扩张和延伸负一层吧台区的实木台面, 使之衍生成为楼梯其中一个踏步, 产生有趣的功能性视觉美学, 让空间和人的行为产生联系和共鸣, 相互协调与统一。

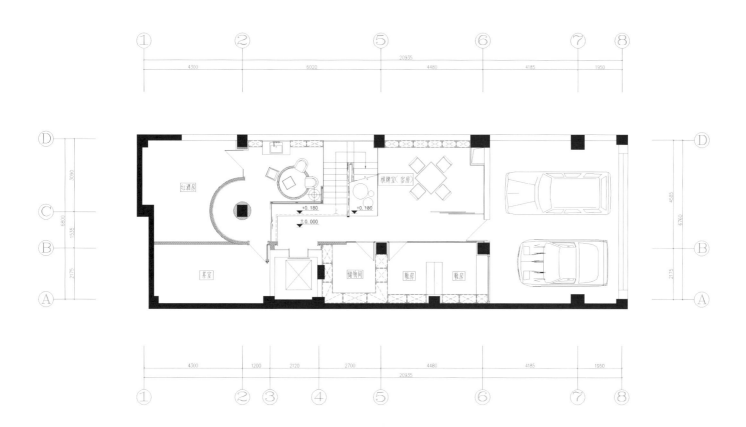

公园1903蒙塔溪22幢负二层平面布置方案图
PLAN 1:100

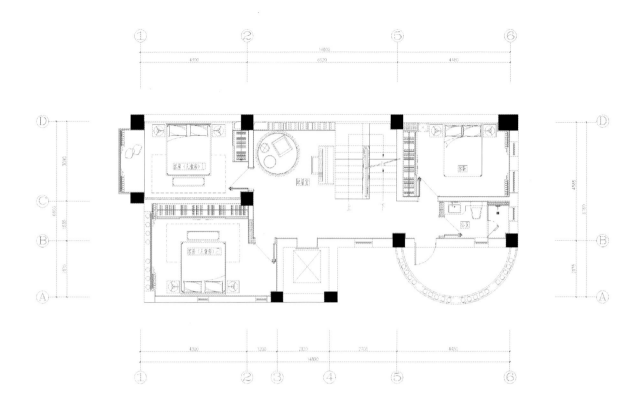

公园1903蒙塔溪22幢二层平面布置方案图
PLAN 1:100

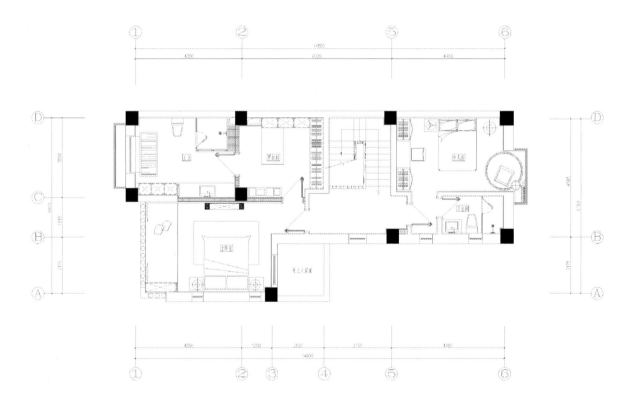

公园1903蒙塔溪22幢三层平面布置方案图
PLAN 1:100

Humanity Gold Award ·
Modern Design Award
现代设计奖·人文金奖

Chen Yang   Zhang Bin

陈洋   章斌

现代设计奖·人文金奖

Design Motto:

A designer should be good at exploring the possibilities of a space with modern techniques and structured blocks, expressing the essence of design by conciliating the conflicts between appearance and inherence, telling stories through design and dedicating to create perfect design works. His thought is a combination of sensibility and rationality and design is an expression of duplicity.

Founder of Between Design Studio
Advocating the variety and innovation of design, his team has completed numerous design projects all over the country.
2018 National TOP 36 for Chinese Design Elites
Member of Institute of Interior Design of Architectural Society of China
2017 Best Residence Award, the 11th DAKIN "New Life Style" Interior Design Competition
2016 Annual Outstanding Home Decoration Work, Jingtang Prize

创意格言:
擅长用现代、块面结构等美学手法演绎和开发空间的可能性,以表像和内在的矛盾统一来表现设计的本质。设计师思维是感性与理性的结合,设计工作实际是表里不一的呈现。喜欢用设计讲故事,并执着于将设计演绎到极致,屡创精彩的设计佳作。

Between 之间设计事务所 创始人
主张设计的多变性和创新性,在全国各地完成了众多设计项目。

曾获荣誉:
2018 年中国设计星全国 36 强
中国建筑学会室内设计分会会员
2017 年第十一届大金内装设计大赛最佳人居奖
2016 入围金堂奖年度家装优秀作品

Recall to Mind

知忆

设计机构：Between 之间设计
主案设计师：陈洋 & 章斌
软装设计师：刘梦 & 小脆
地点：福建龙岩·阳光城
面积：360 m²
摄影：SKY

When living in a space full of various excessive material items, we tend to miss the spiritual nature. We should remove all the false, superficial and useless things, and keep the real and essential things, so as to leave more space and then improve comfortableness and visual beauty.

We deeply believe that, natural beauty without any decoration will never be out of date, and it has constant value rooted in our hearts. We also try to keep distance with vulgar taste and create good taste and style with the time that other people spend on chasing showy and luxury life.

　　当人们的空间被各种物质挤压的时候，也就失去了本质，我们要去掉一切虚假的、表面的、无用的东西，而剩下真实的、本质的、必不可少的东西，从而余出更多的空间，提高舒适度以及增加视觉美感。

　　我们深信，不加装饰的自然之美是永远不会过时的，也是根植你内心的永恒价值。我们努力与世俗的趣味保持距离，把别人追逐浮华的时间用来营造品味和格调。

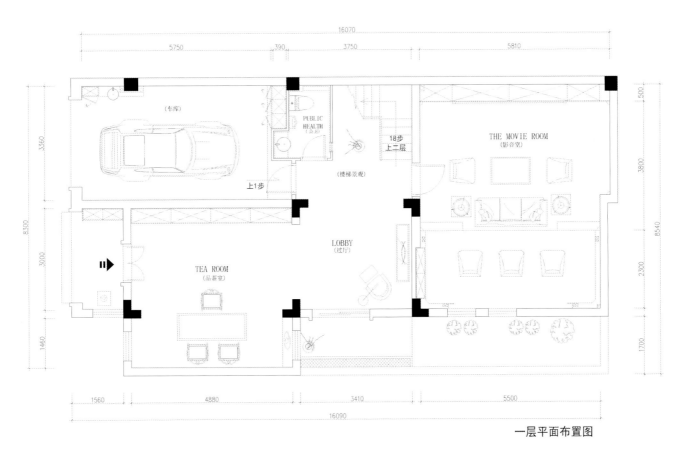

一层平面布置图

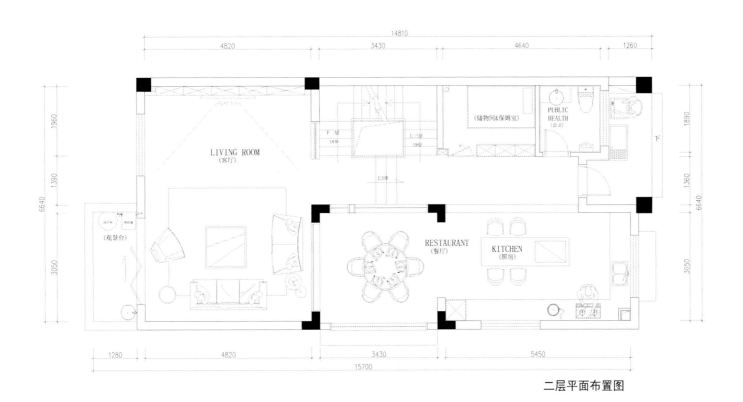

二层平面布置图

三层平面布置图

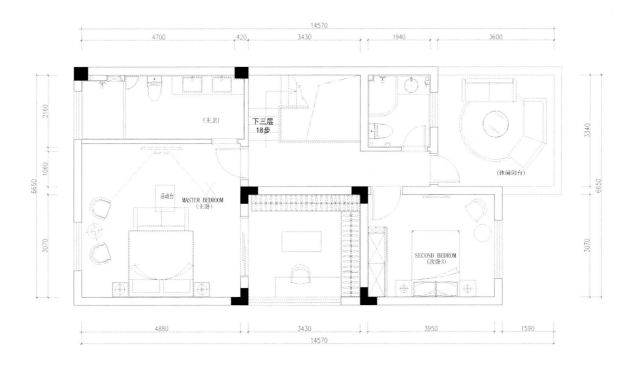

Mirror-like

如镜

设计机构：Between 之间设计
主案设计师：陈洋 & 章斌
软装设计师：刘梦 & 小脆
地点：福建龙岩·卧龙小区
面积：122 m²
摄影：SKY

I keep chasing light because it can turn ordinary things into miracles.

On the basis of guaranteeing the owner's quality life, functionality and practicability are also paid great attention to create unique life experience in this space.

我不断追逐着光，光能将平凡的东西化为神奇。

在满足业主品质生活的基础上，注重功能性与实用性的相辅相成，让空间变成美好独特的生活体验。

Zijinshan Rose Garden

紫金山玫瑰园

设计机构：Between 之间设计
主案设计师：陈洋 & 章斌
软装设计师：刘梦 & 小脆
地点：福建龙岩·紫金山玫瑰园
面积：380 m²
摄影：SKY

As a private residence, the whole space has been given ultimate openness and freedom to avoid a feeling of living in a box. On the basis of meeting daily use requirements, the house type has also been maximized to allow optimum natural lighting condition.

　　对于私宅而言，我们尽量让整个空间给人的感受更加开放和自由，让业主不会觉得是住在一个盒子里。在满足日常使用必备功能的基础上，将户型尽可能改造到最大化，并且实现最佳采光。

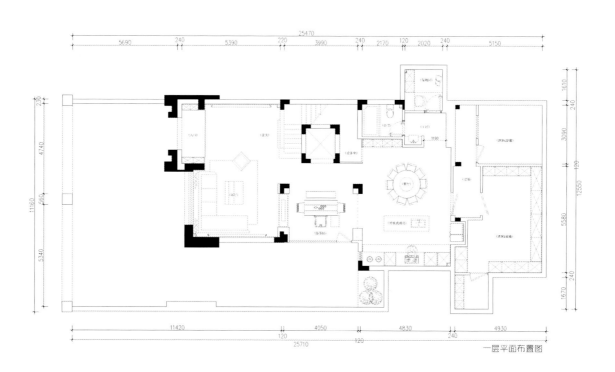

一层平面布置图

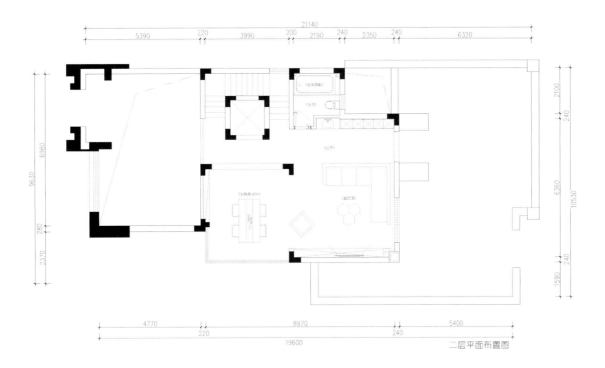

二层平面布置图

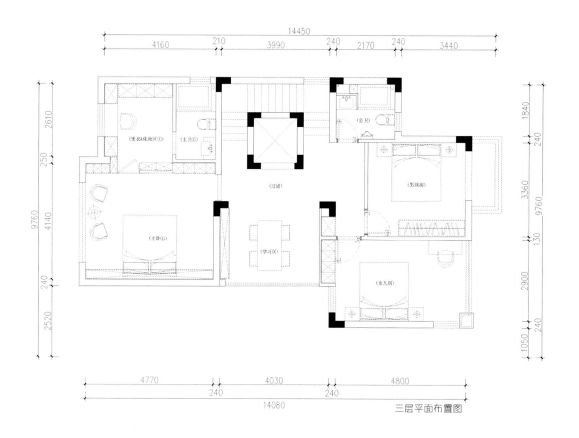

三层平面布置图

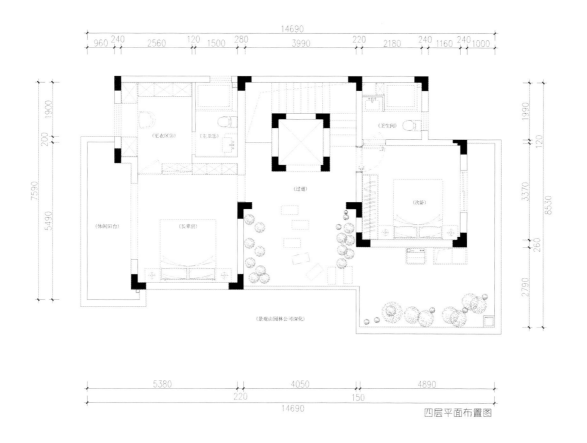

四层平面布置图

Color Scheme Gold Award ·
Modern Design Award
现代设计奖·色彩金奖

Zhu
Lei

朱磊

现代设计奖 · 色彩金奖

2019 Color Scheme Gold Prize, Golden Bund Award
2019 Excellence Award, Best Leisure and Entertainment
Space, Golden Bund Award
2018 China Interior Design Award, KAPOK PRIZE
2018 Silver Prize, Commercial Space, GPDP AWARD (France)
2018 Excellence Award, Leisure and Entertainment Space,
Jingtang Prize

曾获荣誉：
2019 金外滩奖色彩金奖
2019 金外滩奖最佳休闲娱乐空间优秀奖
2018 红棉中国室内设计奖
2018 法国双面神奖商业空间银奖
2018 金堂奖休闲娱乐空间优秀奖

SUMEIMEI

SUMEIMEI 美学空间

设计机构：泓焕时代设计
主案设计师：朱磊
参与设计：武琳
地点：烟台市上市里街区
面积：260 m²
主要材料：艺术涂料、地坪漆、不锈钢
摄影：晟苏建筑摄影

Store image. 店面外观。

SUMEIMEI 美学空间

SUMEIMEI is committed to establish a cutting-edge cosmetic brand image, which covers services like beauty care, body care, manicure and makeup. It is our first time to give it an overall image design and we hope it is simple and new. So its functional areas are simply divided and then connected by colors to highlight women's tenderness and femininity.

Original walls are redesigned with the combination of bright colors and white panels to create a vibrant space. Each room has its own unique color. However, the white hallway has moderated the sharp contrast of color between functional zones, at the same time, connected all the bright spaces. Instead of the bustling surroundings of today's nail salons, our designer hopes to recreate a fun atmosphere. And the interspersing geometric elements, decorative colorful walls and custom chairs like tops are in lively and interesting colors like pink, yellow, red and blue.

Flat and cambered walls divide the whole space into various functional areas, including manicure, beauty care, reception and warehouse, to meet requirements for open or private spaces and enhance customers' experience. Furthermore, the curved mirror frame covers each wall to soften the space, and the circular wall lamp gives one the impression of beautifying themselves in a Hollywood dressing room.

感知色彩在空间中的力量，
少是精简而非空白，
多是完美而非拥挤，
即摒弃不必要的元素，
以呈现事物的本质之美。

SUMEIMEI 致力于成为尖端的美容品牌，服务内容涵盖美容美体、美甲、化妆品。第一次接触 SUMEIMEI 它的产品整体的 VI 设计，我们希望设计更简单化，功能布局更纯粹，通过色彩来连接各个空间，突显柔软和女性的特质，以重塑美容品牌店的形象。

Differentiate spaces by colors.　以色彩来划分空间。

The white hallway has moderated the sharp color contrast.　白色的走廊平衡了彩色空间的强烈特征。

明艳的色彩和白色表面被加以巧妙运用在设计上，原有的墙壁也得到充分利用，转变为一系列富有生气的空间。每个房间分别拥有各自的代表色，白色的走廊平衡了彩色空间的强烈特征，同时将众多明亮的空间连接在一起。设计师希望摒除当下美甲店常有的喧杂环境，转而唤起一种充满探索与乐趣的氛围。几何元素不时穿插在空间之中，带有装饰性的彩色墙面和定制的陀螺椅家具，带来活泼缤纷的色彩，粉色、黄色、红色及蓝色相映成趣。

设计采用了直线与弧线结合的墙面，重新整合了室内界面。弧形墙面划分出美甲、美容、接待、库房等不同的功能区域，满足空间从开放到私密的不同需求，多样的尺度和形状也丰富顾客们的环境感受。打磨成弧形的镜面四角，从细节处强化了室内环境的柔美。圆形壁灯的布置不禁让人联想到好莱坞的化妆间。

A space full of visual impact.
富有视觉冲击力的室内空间。

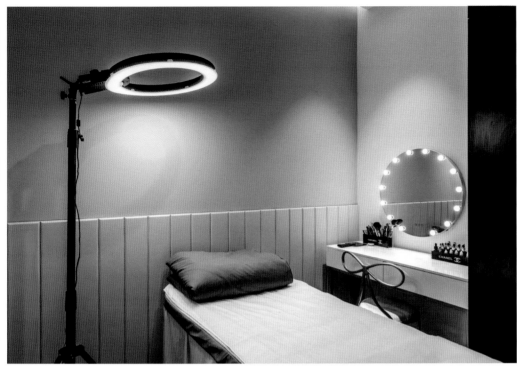

Our designer has specially created a modular menu system (one-stop solution) to show nail polish, colors and pattern, etc.

设计师为美甲店打造了一个模块化的菜单系统，对各种各样的颜料、色彩和美甲款式进行一站式的展示。

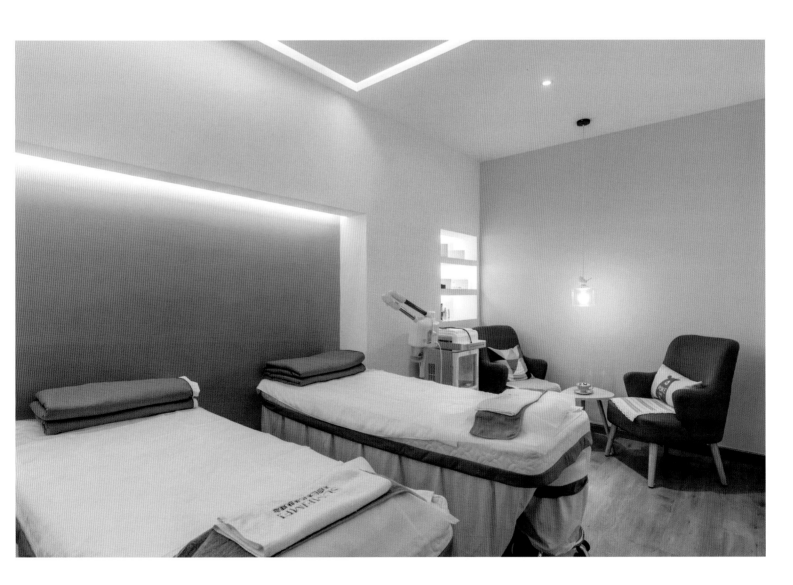

Modern Design Silver Award

现代设计奖·银奖

Shi
Chengju

Wu
Jianbing

师诚举　伍建兵

现代设计奖·银奖

师诚举　　　　　　　伍建兵

2017 Gold Prize, Project Category, Golden Interior Design Awards
2016 SIDA16 AWARD, Singapore Interior Design Awards

曾获荣誉：
2017 年金装奖工程类金奖
2016 年 SIDA16 AWARD 新加坡国际室内设计荣誉大奖

Jianshui Southern Guesthouse

建水南方民宿

主案设计师：师诚举、伍建兵

Jianshui is a famous historical and cultural city with rich cultural deposits and beautiful scenery, which is known as the "famous state of literatures". And its speciality — purple pottery, is listed as one of the four famous ceramics in China.

This case, developed by the real estate developer, is located at one of the entrances of this ancient city. It is a commercial center of traditional quadrangle courtyard style. With a total area of 1,600 m², it has 20 guest rooms arranged on four floors and is also equipped with elevator, staircase, garden, atrium and roof sorden, etc. Moreover, there is a railway station not far away, only 500 meters, so it is very convenient.

The owner prefers a space that combines modern and traditional style, which is consistent with our guesthouse design concept. Therefore, this space is of both modern conciseness and traditional cultural essence, integrated with human geography.

Suitable for the fire-fighting requirements, we add a modern staircase in the atrium, which is so distinct that forms a sharp contrast with the building itself. Texture paint is massively used in the lobby, coffee bar, garden and other areas of the first floor. However, the white wall is decorated with plain stone materials, gray bricks, antique finish and log to create a natural and pure atmosphere.

Spatial pattern of the second floor is similar to that of the third floor. The space between the hallway and the pitched roof on the third floor is fully utilized to create two family rooms. But on the fourth floor, part of the roof garden is transformed into a deck. One distinct feature of this space is that the whole space is well distributed and exploited. A newly-added steel-framed glass staircase leading customers to the terrace on the fourth floor, which, together with the gray tile sloping roof, echoes the theme of the project.

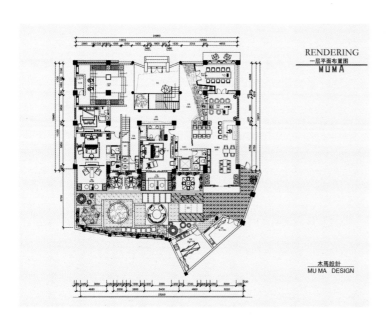

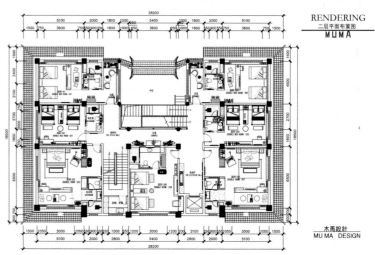

The whole space is properly divided, for example, some guestrooms are inserted with tatami tea table spaces. And every detail is well expressed by simple lines, pure colors and minimalist concept.

In terms of furnishing, we add some ceramic accessories to feature local culture because Jianshui is famous for purple ceramics. The wood furniture also livens the space up.

建水是一个国家级历史文化名城，历史悠久，文化底蕴丰富，风景优美，有着"文献名邦"之称，当地的特产建水紫陶被列为中国四大名陶之一。

本案地处古城其中一个入口处，是由地产开发商统一开发的，模仿古建的一个商业中心，独栋四合院式建筑，项目总面积为 1,600 m²，建筑共有四层，里面有 20 间客房，配有电梯、楼梯、花园、中庭、屋顶花园等基础配备。周边有建水古城小火车站，距离古城朝阳门 500 m，地理位置优越，交通方便。

业主喜欢现代和传统相结合，和我们想做的民宿的概念不谋而合，整体空间简洁现代，与传统结合，与人文地理相融。

因消防要求，我们在中庭位置增加了一把楼梯，该楼梯现代化风格极其明显，与古建相互碰撞，与自然结合，打破常规，形成强烈的对比。

一层的大堂空间、咖啡吧、茶吧、花园等区域，大面积使用了肌理涂料。留白的墙面，加上一些纯朴的石材、青砖、做旧的地坪漆和原木，还原纯粹的质感。

二层与三层的基本空间格局一样，三层靠中庭的两个房间，利用过道和建筑斜顶的空间范围，被打造成了家庭亲子房间。四层因电梯位置，把其中一部分屋顶花园隔到了露台内，空间得到充分利用。新加的钢架玻璃楼梯一直到四层露台的空间，与周围的青瓦斜顶共同突出了本案的主题。

对于房间的设计，我们秉持的理念是随心，随想，随愿，随情。在空间设计上合理划分内部的每个区域，使用舒适简单的线条、素雅的色彩、极简的概念去呈现每一个细节。部分客房通过空间结构划分，加入榻榻米茶台空间，在这里享受冬日的阳光，泡上一杯普洱，看着窗外的景象，不禁让人浅吟"吾心安处是南方"。

在软装配饰方面，因为建水盛产紫陶，我们便加入了一些陶艺的配饰，作为和当地文化的融合。家具的选择上采用了原木色搭配，让整个空间充满生机。

"随心，随性，拥有南方的温度，吾心安处，让我们遇见你"，是南方民宿的理念，也是我们的设计初衷。木马设计，只做有温度的设计。

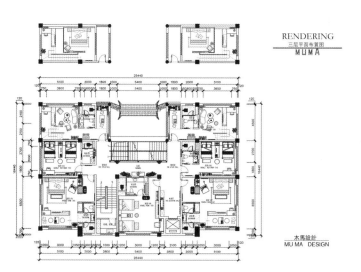

木馬設計
MU MA DESIGN

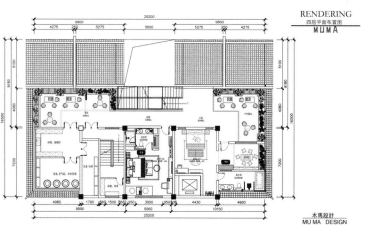

木馬設計
MU MA DESIGN

随心随性

拥有南方的温度

吾心安处

让我们遇见你

Li
Jie

李杰

现代设计奖 · 银奖

Exploring the possibilities of space persistently and showing ultimate details without meanness.

2018 Silver Prize, Tuozhe Cup — Residential Space Design Competition
2018 Niceliving Awards
2018 National Top 100 Interior Design Talents

执着于探寻空间之衍，
不吝靡费时间也要呈现细节所到之处。

曾获荣誉：
2018 "拓者杯" 居住空间设计银奖
2018 好好住营造家奖
2018 达人室内设计全国百强

Helen's Spring

海伦春天

设计机构：西安素图空间设计
主案设计师：李杰
面积：70 m²

The project is the standard dwelling size with two rooms, one living-room, one kitchen and one bathroom, covering an area of 70 m². Its main color scheme is black and white, and a little red.

As a function area, the living room is decorated with black and white, and then a small amount of red ornaments are added to naturally constitute an art piece, which echoes with the painting on the wall behind the sofa.

Limited by the dwelling size, the living room and dining room are put into one open area. The same tone — black and white, echoes with the metal droplight, which expresses a sense of quality. Considering the space for storage needs, our designer chooses a set of cabinets opposite the entrance. Mirror effect of the cabinet surface and layering floor texture extend the space visually.

与这黑白交际线上行走，设计仿佛有一种绮丽的魔力，于虚无之中留下一抹赤色，去融合这黑与白。

所有的相遇都归结一个方程式"六度连接"，即朋友的朋友的朋友就是朋友。

设计亦复如是，我们和业主的相遇也可以说是一次连接，业主事先找到的是广东那边的"射鸡狮"，后来经过介绍，我们和业主便开始了这场"春风十里，把家给你"的交托之旅。

　　经了解，业主是一位 88 年出生的祖籍四川人，他们购置的房子是标准的两室一厅一厨一卫户型，面积为 70 m²，在设计上主要运用了简约的黑与白作为整个户型的主色调，少量的红色点缀其中，在满足业主品质生活的基础上，又注重功能与实用性的考量，像这黑与白的融合一般迸发出无限的可能。

　　客厅作为起居室的应用，在这里其功能与美都得到了极致的显现，黑与白之间运用少量的红点缀其中，构成了一幅浑然天成的艺术画，和沙发背景墙上的艺术挂画相得益彰。赤墨映雪睹梅香，闻香客来满帝都。

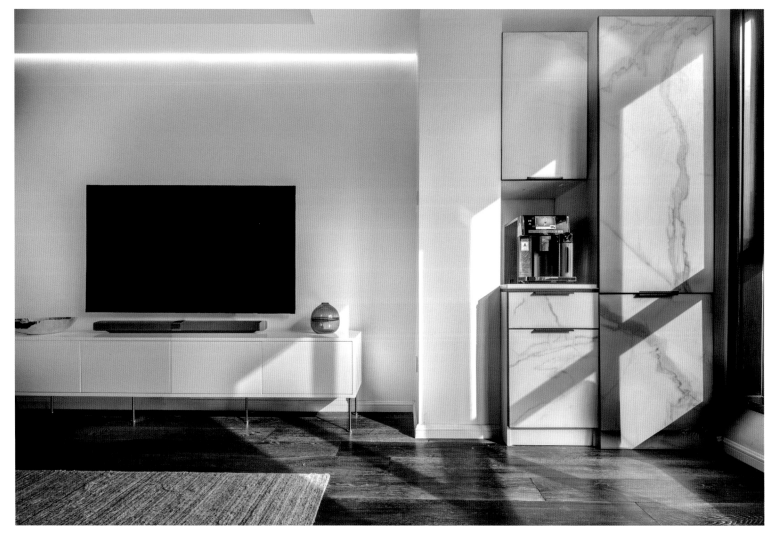

由于户型的关系，客餐一体
是唯一的选择。餐厅空间沿用了
黑与白的主基调，搭配金属吊灯，
品质与质感得以彰显，考虑到收
纳空间的需要，在入户门对面做
了一组柜子，柜门运用了镜面材
质，以及地面地板的铺设，顺从
纹理的方向铺设，从视觉和心理
上延伸空间。

The L-shaped kitchen is well fitted into this space, which can meet all kinds of needs: washing, cutting and frying. We want to improve the daylight and improve the interaction among family members, so the wall beside the door is replaced by a glass door.

As for the bathroom, it follows the same tone. An invisible sliding door makes the space fun.

　　L 形的厨房与这空间配比如此契合，在满足洗切炒的功能需求上，我们在设计时想让厨房里的主人公与家人有所联系。所以，我们将厨房门一侧的墙推掉，做成玻璃材质的门，既增加了家人之间的互动，又增加了厨房的采光。

　　卫生间的设计和整个空间基调相统一，隐形门的设计，在开合之间玩味着空间的趣味性。

In the children's room, a bunk bed is the best choice due to limited space. Shades of green and other jumpy colors are added to the black, white and gray space to create a vibrant scene.

　　儿童房上下铺的选择是既定空间的限定，也是空间提升利用的主选，黑白灰之间注入草木绿和少量的跳跃色，使得整个空间充满生气。

Jiu Jin Tai

九锦台

设计机构：素图设计公司
主案设计师：李杰

In this case, out of respect to the original building, structure and layout haven't been greatly changed. The space's main color — white, and the using of the wooden texture make people feel warm. The wallcovering with delicate texture has been applied to walls and ceiling, which enriches our sensory experiences.

Because the original structure provides no transit space from outdoor to indoor, our designer leaves a small area at the entrance for a porch shoe ark and a red sofa, which is convenient for the owner to change shoes.

With careful thoughts, there is no visible partition in the sitting and dining area, only a passageway between them, so they are a whole. What's more, a part of the wall in the sitting room has been removed to design a hidden door which is flush with the door of the master room. The recess of the wall opposite to the master bedroom is turned into a locker. And wood panels extend all the way to the background wall in the sofa area, which is a skillful design to abandon redundant furnishings.

本案在空间上并没有进行大刀阔斧的改造，而是在尊重建筑物的结构本体做了局部的微调，整体运用白色作为基础色调，木质材质融入空间之中，给予整个空间一种温暖的感觉，墙面和顶面采用了海基布，其材质细小的纹理带给整个空间一种特别的肌理感。

原始的户型结构并没有为入户玄关的设计提供较好的空间基础。为了满足业主一家人的生活需要，设计师在进门区域处预留了玄关柜的位置，考虑到业主换鞋的便利性，配了一把红色的座椅。

客餐厅区域在空间上没有明确的划分，以过道为媒介自然地将两个区域做了软隔离，进门看去客餐厅是一体的，这就要归功于设计上的一些小心思，设计师从墙面着手，将客厅区域的墙体拆除掉一段，和主卧门垛平齐做了一樘隐形门，利用主卧对面的内凹的墙体结构，做了一组柜子以提供储物功能，木饰面板从餐厅一直延伸到沙发背景墙，这种大块面的设计简化了多余的、繁杂的片段，阳光越过窗玻璃散落斑斑驳驳的光影，风轻轻摇晃窗外的琼枝，静待流年时光。

The kitchen is designed as a U-shaped space which equipped with a refrigerator. The moving direction, fetch — put — wash — cut, is reasonable and clear. The corner between the balcony and study is upgraded into a tea-making and tea-tasting area.

厨房被设计成 U 字型的空间，并且将冰箱容纳入其中，取 — 放 — 洗 — 切的动线分布合理。阳台与书房相接的一角被设计成品茶的区域，屋主阳台上种植的多肉宝宝们沐浴在阳光里，茁壮成长。

Originally, there is only one window for the study, but the designer removes the wall and makes it into a door to connect the study and balcony. And then the wall on the left of the entrance is turned into the wardrobe for storage.

书房原本是有一面窗户的，我们设计师打掉了窗户那面墙，留出一个门垛，使这个空间和阳台空间联系起来，然后利用进门左边门垛一面墙做了整面墙的柜子，来满足收纳空间的需求。

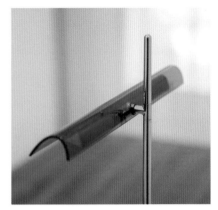

As for the bedroom, designer makes no change to its layout, but the whole space is well handled. The background wall of the bed follows the same hues of the wood panels. Two different colors from top to bottom make this space more distinctive.

设计师没有对卧房的格局做任何改造，只是把握一个尺度关系，主卧的床头背景沿用了和整体色调一致的木饰面，这种上下分色的处理，和卧室其他空间区分开来。

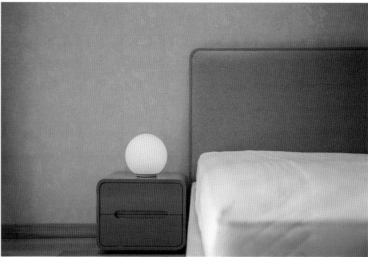

Laian Cozy Space — Light Chasing
莱安逸境·捕光

设计机构：素图设计公司
主案设计师：李杰
面积：89 m²

This project is designed for a family of three, the porch is on the left side of the entrance, but the space is too limited to meet each member's requirements for storage, so designer removes the left wall to make a wall of cabinets. Users can press to open the doors, which is very convenient.

Part of the wall of the balcony has been removed, which enables the sunshine to light up the whole dim space. The visual expansion is expended by connecting the living-room with the dining-room. The furnishings and the whole space are white or black to create the sharp contrast, and then embellish a slight warm yellow.

经了解，业主是俩夫妻，有一个小男孩儿。89 m² 的空间里，三口之家的故事开始在我们素图空间设计中演绎一场捕光之旅。

进入业主家量房的时候，我们发现他们家的采光特别差。进门右边厨房和卫生间紧密相连，客厅阳台区域立了一堵墙，开了一个小小的门洞，客厅窗户的光照引不进客厅区域。在设计上我们以极简风格为主轴，在空间里做减法，强调黑白之间的极致碰撞，从视觉上拉伸空间感。

一进门左边是玄关区域，考虑到要满足一家三口的日常收纳需求，进门的区域预留玄关尺寸不够，因此设计师把左边墙体也全部砸掉，做了一整面的玄关柜。柜门采用按压式，使用起来更方便。

至于客厅，在经过仔细的考量之后，设计师打掉了阳台区域的墙体，使得整个昏暗的空间暴露在阳光下，客餐厅合为一体，视觉感官上得到拉伸，整个空间包括后期的家具配饰选择，都是以一种极致的黑白来拉开对比，运用少量的暖黄灰做点缀处理。

Part of the TV wall between the sitting room and the master bedroom is removed to add sliding glass doors, which is a good way to transform the original space.

对于主人房和客厅电视墙，设计师打通了部分墙体，做了玻璃移动门，使得原有的空间限定不是很鲜明。黑白灰的调性，加上剔透的玻璃，光影散落，自然闲适。光与影之间，你的起落坐卧都在我的眼里。

The kitchen and bathroom are connected closely without enough natural light, so the wall between the kitchen and bathroom on the right side of the entrance is removed, too. Here, transparent Changhong glass is adopted to improve lighting condition.

Bathroom is like a long rectangular box. All the sanitary wares are arranged in a line, so the space is spacious vertically. The L-shaped kitchen can meet all needs: washing, cutting and frying. The double-door refrigerator is also put into the kitchen to make life more convenient. Moreover, glass doors break the fixed form of the space.

　　厨房和卫生间紧密相连，且采光条件不好，为解决光线问题设计师进行了前期勘测划线定位，进门右边厨房和卫生间的墙体是可以拆除的，因此设计师拆掉了这块儿的墙体，改用长虹玻璃来增加透光性，把光引进厨房和卫生间。

　　卫生间是一个狭长的矩形盒子，设计师以一字型的设计，让所有的使用构件都保持在同一水平线上，在立体空间上，以竖向掏砍的方式来满足卫生间的功能需求。L形的厨房设计满足了洗切炒的功能刚需，双开门大冰箱，使用起来更加便利。玻璃门的设计弱化了固定的空间界定形态，增强了女主人和家人的互动和联系。

At last, the children's bedroom has a wall of wardrobe and desk to improve storage capacity.

　　儿童房的主人是一个小男孩儿，钢铁侠和蜘蛛侠是他的最爱。和大多数的家庭一样，孩子所需要的收纳储物是最让人头疼的，所以设计师留了一面墙，将书桌衣柜集于一体。

Ren
Zhehui

任喆辉

现代设计奖 · 银奖

2018 Best Residential Apartment Design Award, Jingtang Prize
2018 the 3rd "BERYL CUP" International High-end Residential Design Award
2018 Apartment Design Award, the 8th "ISD Award"
2018 Bronze Prize, Apartment Design Award, the 9th IDEA-TOPS

曾获荣誉：
2018 年金堂奖年度优秀住宅公寓设计奖
2018 年第三届"BERYL 杯"国际高端住宅设计奖
2018 年第八届"IDS 大奖"公寓空间设计奖
2018 年第九届 IDEA-TOPS 艾特奖公寓设计奖 铜奖

Concise and Delicate
简一致

设计机构：台州市丽园装饰公司
主案设计师：任喆辉
面积：228 m²
地点：台州江南首府
主要材料：意大利灰大理石、胡桃木木饰面、灰色艺术涂料、实木地板等

This is a simplified Taiwan-style project. The walls are decorated with no skirting lines, so the whole space is concise. Doorframes are narrowed to 4 cm; wardrobes are frameless; and built-in air conditioning outlets are upgraded to stainless steel. All these details make this space simple and generous.

Openness is the main design concept for the whole space. As for the layout, each space is maximized to ensure transparency, for example, half of the balcony becomes a part of the living room, which makes the whole space more comfortable and spacious. Moreover, the living room, dining room and kitchen are arranged on the same side along the axis, which facilitates the transparency and integrity of this function area. In the master suite, a wardrobe hidden in the wall provides abundant storage space, which gives more transparency and brightness to the study.

Rich materials are applied to this space, such as walnut, Italian gray marble, imported wall coating and dark grey wood floor. The warm feel and texture of walnut matches well with the coolness of gray marble; the light gray walls and dark gray floor make the space a slight jumpy. Color of the droplight is similar to the chairs. And marble in the kitchen is of the natural texture.

本案整体风格上以台式为主，在台式的基础上进行简化，比如全屋采取无踢脚线的处理方式，使空间显得更加干净利落，全屋门套采取 4 cm 窄边处理，全屋衣柜柜体采取无边框式处理，全屋空调出风口采取内嵌式不锈钢风口方式，使得整体空间更加简化，更加干净。

整体空间上采用开放式为主的设计，空间布局上的主线是把每个空间尽量做大，视觉感尽量无阻隔。比如，客厅外阳台内包进客厅，然后客厅适当外移内包阳台空间，客厅空间更加舒适，同时餐厅空间更加宽敞。继而成客厅餐厅厨房在一个轴线上一字排开，显得整个空间在通透中又是融合的。取消主卧衣帽间的空间，将套间式书房的一面墙全部处理成隐形衣柜，在保证储藏量的前提下把衣帽间的空间腾出来，使得套间式书房的空间显得更加通透明亮。

In the study, the same walnut, wall paint and dark gray wood floor as those in living-room are also adopted to produce a natural and simple atmosphere.

In the master bedroom, walnut, hard padding and gray wall paint are well matched to make the space peaceful. Two small droplights over the bedside are simple and graceful. However, the daughter's room is white and pink, which makes the space lively and full of childlike fun.

该项目全屋选材上选用了山纹胡桃木、意大利灰大理石、进口墙面涂料、深灰色大板地板等材料。胡桃木的手感和质感都是十分细腻和舒适的，有人体的温度，结合少量的灰色大理石，温度中带少许冷冷的酷感，墙面浅灰和地面深灰显得空间带有跳跃性。餐厅吊灯选择了和座椅接近的色调，暖光静静地洒下来营造出满满的温馨感。厨房利用大理石自然的纹理进行对比、融合，简约的厨房空间在这样的氛围下诞生的一道道美味，是熟悉的家的味道。

书房空间运用了与客厅一致的山纹胡桃木材质、深灰色大板地板以及艺术涂料，无论是闲暇的午后，还是宁静的夜晚，书房总是填补空缺的港湾，卸下整天工作的疲惫，阅一本好书，赏一部电影，悠然自得。山纹胡桃木护墙元素的运用，造型简约的书桌书柜，让书房多了一份返璞归真的气质，心境更加豁达高远。

主卧室采用了山纹胡桃木及硬包材质，让空间显得更安静，一间舒适的卧室，必然能够为人带来身体和精神的松弛，直达内心柔软之处。墙面艺术涂料灰色调铺陈安静，床头的小吊灯造型简约大气。女儿房用粉色搭配白色打造出了更加轻快活泼富有童趣的氛围。妈妈心中的小公主，自当住在浪漫梦幻的童话王国，轻纱帷幔，每一帧都是成长记忆里的美妙时光。

家，应该是一个能让人自在放松的地方，不管是日常生活还是视觉感受，舒适才是最主要的。在这里，每一个细枝末节都围绕着生活和爱的主题，每一个细节，都让你爱回家。

Time and Space
时光与时空

设计机构：台州市丽园装饰公司
主案设计师：任喆辉
面积：500 m²
地点：台州绿城宁江明月排屋
主要材料：灰色大理石、科技木木饰面、素色壁纸、实木地板等

This case is of modern Zen style. It includes 5 floors: a basement, three floors in the middle and a loft.

In the sitting and dining area, gray stone material matches with wood panels to create a layering effect. And the steel grillwork on the two sides of the staircase in the living room becomes the spotlight of this case. It separates the space unconsciously and also becomes the background of the foyer, which creates a sinuous effect from geomantic theory. Later, the whole space is decorated with furniture and accessories in Zen style to enhance its modernity and Zen mood.

At last, bedrooms are mainly decorated with plain wallpaper and some decorative wood panels to create a quiet and natural atmosphere.

黄岩宁江明月12幢3号排屋一层平面方案图 1:55

本案装饰风格是以现代简约风为主线，再加些禅意风的后期装饰，使得整体空间能达到现代禅意风的视觉效果，也能体现出业主有着艺术底蕴的生活品味。本案一共5层，地下一层，地上三层加顶层阁楼。本案客餐厅采用了灰色调的石材，奢华而不俗套，低调而不平凡。期间，搭配着木饰面使整个空间更加丰富。同时一致的色调，令空间整体感觉非常舒适。客厅铁艺花格设计是此案的一个亮点，花格设计在楼梯两旁，使得一层在整个一体的空间内，通过花格的设计在人的意识中自然形成一个分区，又能做一层入户玄关背景，形成了风水上讲的迂回感，是一举多得的设计。在后期通过禅意风的家具和软装搭配，使整体在现代风的基层上体现出禅意感。

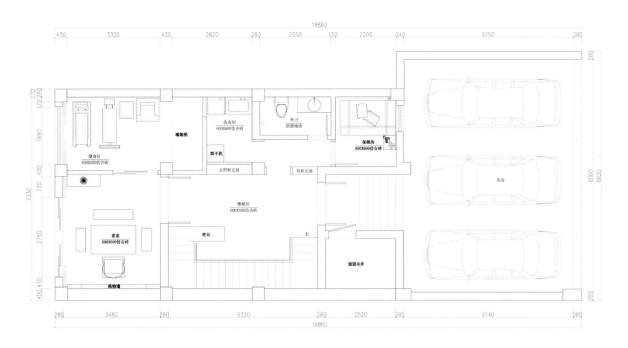

黄岩宁江明月12幢3号排屋地下室平面方案图 1:55

黄岩宁江明月12幢3号排屋二层平面方案图 1:55

卧室设计主要以素色壁纸为主，加上木饰面的点缀处理，使人感觉整体空间能安静下来，让主人回到家进入卧室，能够褪去城市的喧嚣，可以有个较为安静的空间好好休憩，感受自然以及空间设计给自己带来的舒适和放松。

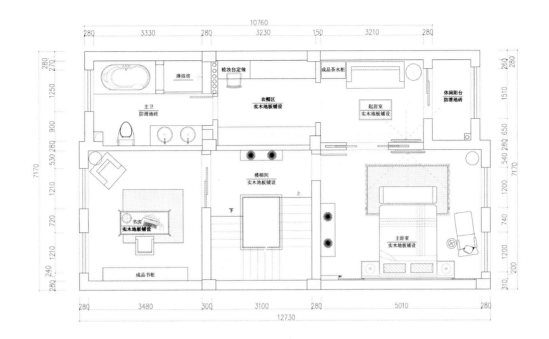

黄岩宁江明月12幢3号排屋三层平面方案图　1:55

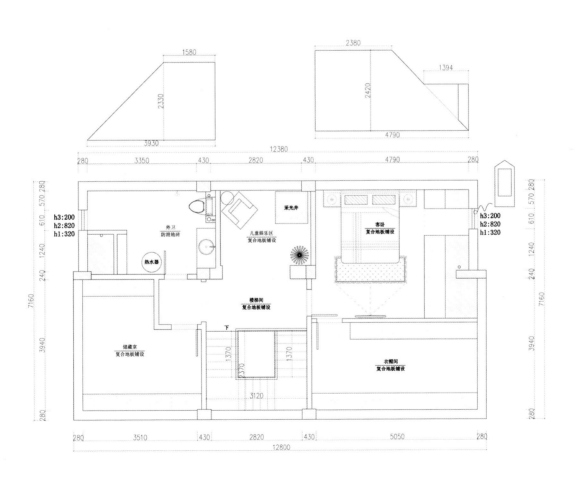

黄岩宁江明月12幢3号排屋四层平面方案图　1:55

Wu
Donglong

吴东龙

现代设计奖·银奖

2017 Elegant Home Prize, the 3rd Xiamen Beautiful
Home Design Competition "Ji Jia Cup"
2016 Elegant Home Prize, SHOKAI Brown Stone Villa
Custom Refined Decoration Design Competition
2015 Cozy Home Prize, the First Xiamen Beautiful
Home Design Competition "Ji Jia Cup"

曾获荣誉：
荣获 2017 "吉家杯" 第三届厦门美宅设计大赛
（雅居奖）
荣获 2016 "首开褐石墅" 定制精装设计大赛
（雅居奖）
荣获 2015 "吉家杯" 厦门首届美宅设计大赛
（尚居奖）

Xinjing International Bund

新景国际外滩

设计机构：厦门东方品位装饰
主案设计师：吴东龙

For this case, the original small dining hall is renovated into a foyer, which is divided into two parts by a partition made of wood battens. But the new dining room is upgraded from a bedroom. And between the living room and dining room, there is a hollow-out wardrobe made by thin steel panels which makes the space transparent. The original bedrooms on the 2nd floor are preserved, but the heightened living room becomes the master room.

As for the foyer, beige and bison wood panels butch with the gray marble floor, which makes here high-end and elegant. What's more, the dining room and the living room are designed with a similar color scheme — white and beige, but a small amount of light blue decorations are also adopted to liven the space up. And its designer also draws the outline of this space with neat modern symmetric lines. However, the imported maple laminate wood floor is only used to highlight the spotlight of this area — the TV wall and the background of the dining room. Light maple hue here endows this space a natural and refreshing atmosphere, but the bison staircase makes the space dignified and sedate.

一层平面布置图
SCALE 1:50

11320

3370　240　3890　170　1880　100　1670

240 410

3220

120

3980

7970

2620

100

7440

3200 ENTRY

1520

1250　620　5220　2110

9200

观景阳台
VIEWING BALCONY

餐厅
DINING ROOM

厨房
KITCHEN

卫生间
REST ROOM

门厅
HALL

客厅
LIVING ROOM

上17级

由于该小区的特殊性和业主极高的鉴赏力，本案空间经历了一次颠覆性的大调整。入户原本的小餐厅被改造成一个入户玄关，给客厅增加了一层私密感，透空的木线条划分两个区域，既有分隔又有空间延伸，让空间与空间有了交流。现在的餐厅则是由原来的房间改造而成，在客厅与餐厅的中间采用了超薄钢板制作而成的透空装饰柜，空间穿透力极强。二层其他卧室全部保留，原本挑空的客厅倒板改成主卧。

玄关采用米灰色与深咖色木饰面与灰色大理石地面搭配，营造出高级优雅的观感。客餐厅整体色调上以白色和米色为主，简洁干练的现代不对称线条勾勒出整个空间，局部淡蓝色调的装饰品突显活力，进口枫木强化地板的电视背景与餐厅背景，强调其空间主体性，清淡的枫木本色赋予了清新自然气息，楼道的深咖色木饰面给清新的空间一份沉稳、安静。

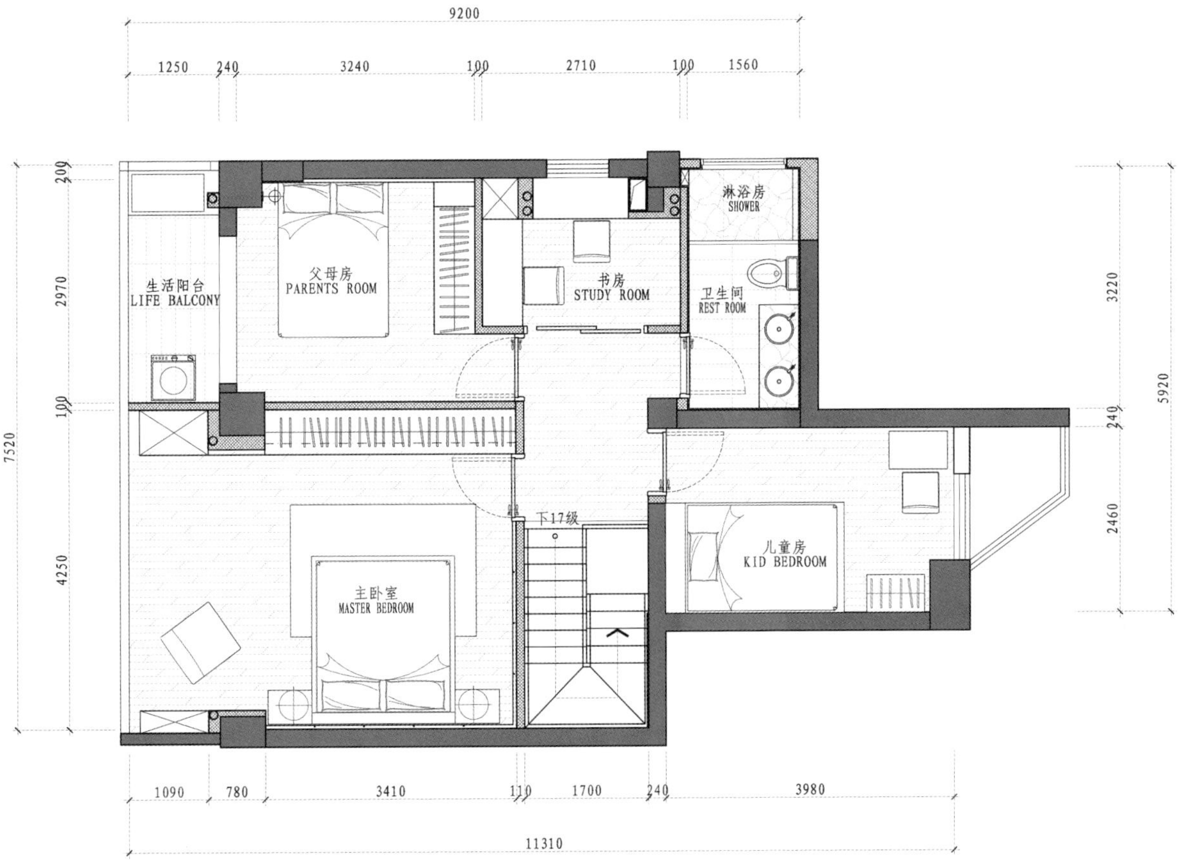
9200
1250
240
3240
100
2710
100
1560
生活阳台
LIFE BALCONY
父母房
PARENTS ROOM
书房
STUDY ROOM
淋浴房
SHOWER
卫生间
REST ROOM
儿童房
KID BEDROOM
主卧室
MASTER BEDROOM
下17级
200
2970
100
4250
7520
3220
5920
240
2460
1090
780
3410
110
1700
240
3980
11310

二层平面布置图

SCALE 1:50

Cai
Fei

蔡飞

现代设计奖·银奖

2012 365 Home Design Competition
2010 Third Prize, China Top 10 Elite Designers Competition,
KITO Cup

曾获荣誉：
2012 年 365 家居设计大赛
2010 年金意陶杯中国十大设计师新锐大奖赛　三等奖

Tanghua-Seiko Exhibition Hall
唐华舒适家展厅

设计机构：观想空间设计
主案设计师：蔡飞
地点：西安市曲江新区

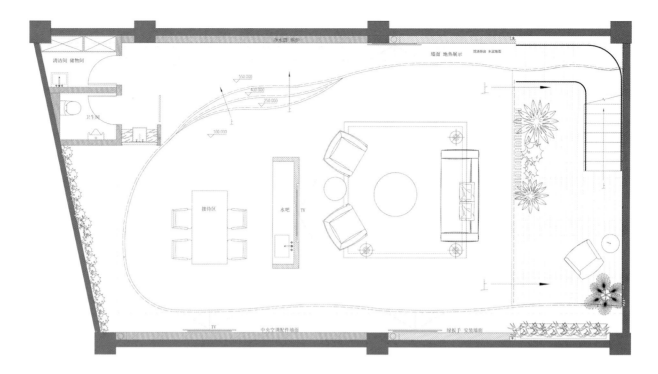

负一层平面布置图 1:80

Tanghua-Seiko is located in the south gate of Qujiang Purple Garden, Xi 'an, which is a living experience center equipped with central air conditioning, air purification, water purification and water heating systems.

The whole space is designed with various scenes of real life, such as the long wooden bar counter, tea room, beer bar and DIY coffee area.

As for the basement, "view borrowing" and "light borrowing" techniques are adopted to bring in natural sceneries through the French window. As an area with various experience levels, the displays in the room also improve ventilation condition in the room.

唐华精工舒适家位于西安曲江紫汀苑南门，是一家集中央空调、新风净化、净水系统、采暖热水四大系统于一体的生活体验馆。走进去，你感受到的不是冰冷的各种型材电器，而是轻松、自在的舒适氛围。

整个空间融入许多生活化气息的场景，例如有超长的木质吧台，有可以品茶的茶室，有精酿啤酒吧区、还有自制咖啡区。

地下室通过"借景""借光"，让空间得以延展，透过落地窗可以欣赏室外的风景。而作为深层次体验区，所展示的产品也有利地证明如何改善房间不通风的问题。

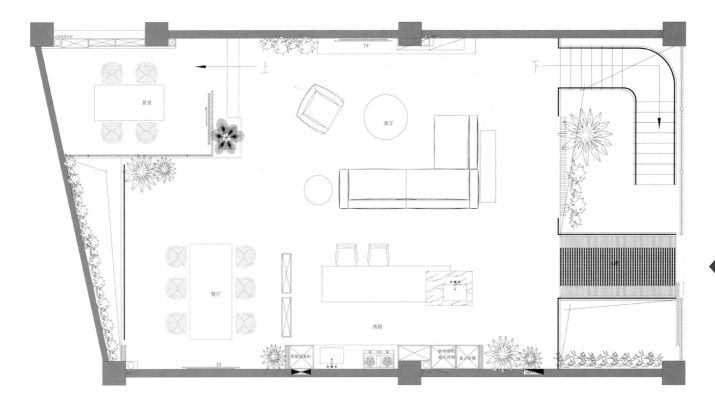

一层平面布置图 1:80

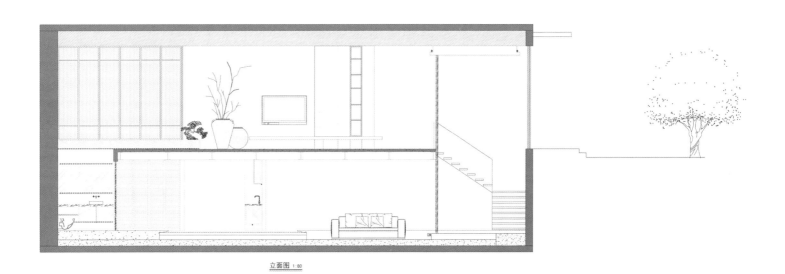

立面图 1:80

Aesthetics Award ·
Modern Design Award
现代设计奖·美学奖

Luo
Nan

骆南

现代设计奖·美学奖

2018 Excellent Designer of China Interior Design Industry
2017 Jin Teng Award, Tencent Home
2014 Annual China Interior Design Award, Jingtang Prize
2013 the 2nd Nest Award
2012 Gold Award, Love Home Design Award

曾获荣誉：
2018 年中国室内装饰行业优秀设计师
2017 年腾讯家居金腾奖
2014 年中国室内设计年度金堂奖
2013 年第二届筑巢大奖
2012 年爱家杯设计金奖

Western Sea View

西派澜岸

设计机构：北京龙发建筑装饰有限公司成都分公司
主案设计师：骆南

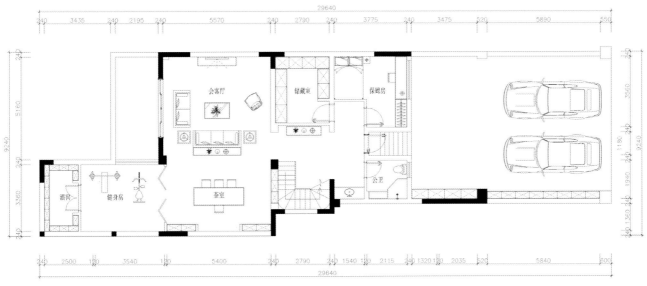

负一楼平面布置图

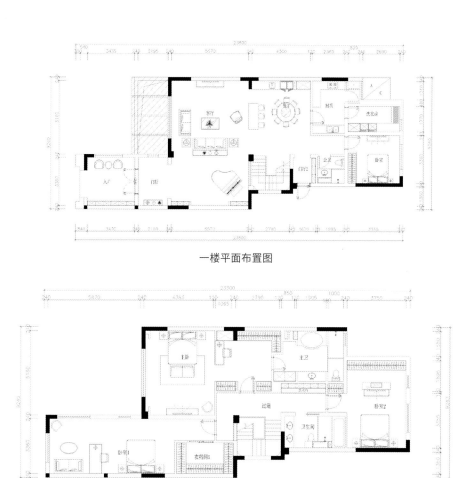

一楼平面布置图

二楼平面布置图

This townhouse has 3 floors with an area of 468 m².

Its designer defines this project as a modern style. A perfect match of cool and warm materials like metal, stone, wood and fabric gives one an elegant and warm feeling. This spacious space also covers various connected function areas to provide a convenient life for the owner. Moreover, living in this space, calmness and elegance can also be felt through the combination of dots, lines and faces as well as dark and light colors.

该项目为叠拼别墅，分为三层，空间面积为 468 m²。

设计师针对本项目选择了现代风格，整体简单大气的氛围体现了主人低调得体的气质。金属与石材这种冰冷的材质结合温柔的木质、暖和柔软的布艺，呈现出高雅温馨的感观。大气的空间，足够体现出该设计多功能的用处，紧密相连的功能区对生活而言再方便不过了。各个空间运用点、线、面的结合，深色和浅色的相呼应，突出空间的稳重和高雅。

Mountain View

观岭

设计机构：北京龙发建筑装饰有限公司成都分公司
主案设计师：骆南
面积：380 m²

This detached villa covers an area of 380 m², including 3 floors.

The whole space is of modern style; its main color scheme is cool and refreshing; and its function areas are comprehensive to meet various requirements. All in all, this elegant and neat space is a dreamland for girls.

此项目为独栋别墅，空间面积为 380 m²，分为三层楼。该房子的主人是一个成功女强人，其家人偶尔回来住。

该项目主人虽然是个女强人，但是也有着自己温柔的一面，回到家中便回归天性，做回小女人。整个空间设计为现代风格，在色彩搭配上主要以清新色调为主。在功能分区上满足主人的各种需求。设计整体给人一种淡雅、简洁的感受，打造出女孩子梦里想要的场景。

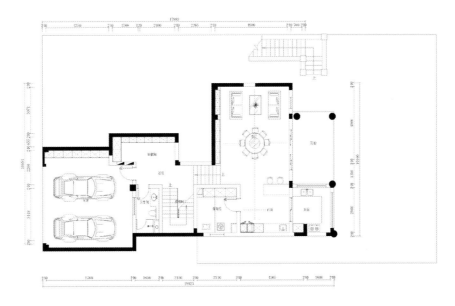

负一楼平面布置图

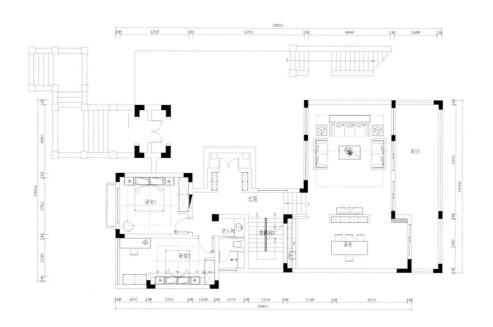

一楼平面布置图

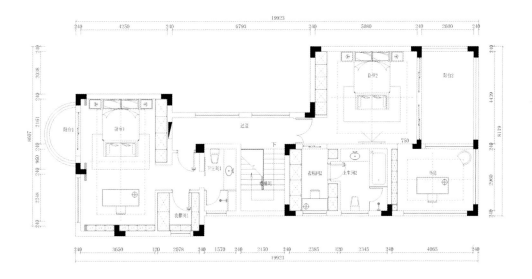

二楼平面布置图

Delicacy and Simplicity

极致简约

设计机构：北京龙发建筑装饰有限公司成都分公司
主案设计师：骆南
面积：800 m²

As for this villa, it has an area of 800 m² with two floors.

The owner of this villa loves a clean and generous space, which is different from most elderly people who like complicated furnishing. Adopting the brief and smooth modernism architectural style, a combination of white walls and wood texture makes the space calm, mature and tranquil. Also, this heightened space gives a sense of spaciousness and its diversified functions can meet every family member's need.

该项目是一幢独立的两层楼别墅，空间面积为 800 m²。

对于老年人来说，更注重家给他们带来的感觉，在花甲之年的日子想要回归轻松的生活。大部分老人喜欢复杂繁重的东西，可主人却更喜欢干净大气。本项目设计采用现代简约的风格，白色与木色的搭配，营造了一种稳重、成熟、安静的气息。挑空的空间彰显出大气，空间功能多样，满足所有家庭成员的需求。

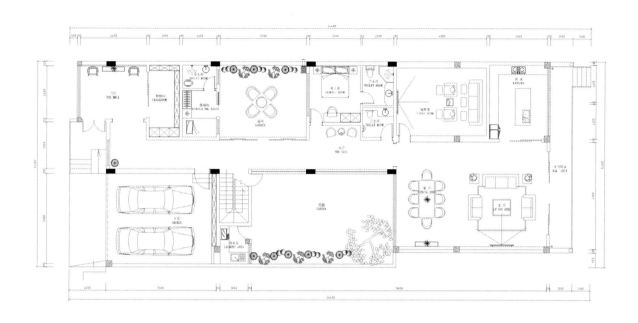

一楼平面布置图

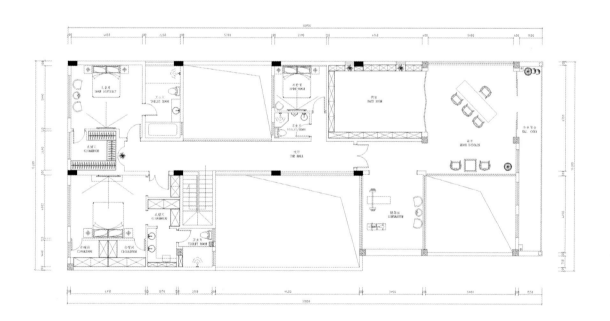

二楼平面布置图

Jin
Xin

金鑫

现代设计奖·美学奖

Office of Da Bin Space Design
大斌设计办公室

设计机构：大斌空间设计
主案设计师：金鑫

This project is designed with diversified functional areas, including a foyer, reception room, water bar, office area, storage room, etc. And all the areas are interwoven vertically or horizontally with a clear flowing direction.

As the main area, working space is symmetrically arranged and fully equipped.

Layout of the working area on the 2nd floor is inspired by Piet Mondrian. Concretely speaking, the ceiling is segmented by linear lights to meet lighting requirement and enhance artistic effect; through partition techniques, noticeable private offices are also created like isolated boxes, which seem to be small buildings in a big building. So the whole space is full of fun.

Stone, wood and stainless steel are skillfully applied to the foyer, working area and water bar to highlight a sense of modernity and innovation.

A 720°spiral staircase connects the first floor to the second floor. Its smooth lines make the whole space artistic.

The reception area is decorated by wood panels and stone material. To connect the whole space, wood texture barriers are arranged vertically and then extend horizontally.

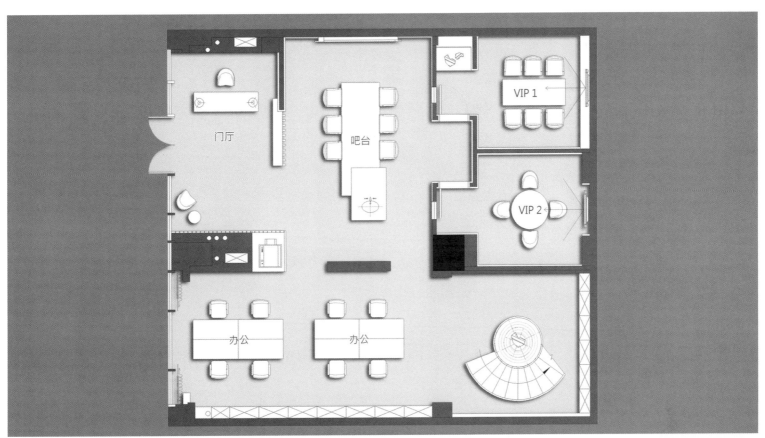

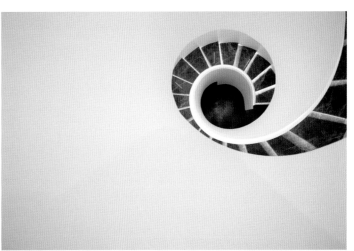

男卫　　女卫

清洁

办公

办公

办公

茶水　　储物间

办公

业务部

办公

DABIN
SPACE DESIGN
MAY 2018

My life can only be meaningful if I can be
willful to insist on what I like.

本次设计改造将为其注入新的功能和活力，为大斌空间设计打造一个全新的、舒适的办公场所。整体平面设计包括了门厅、洽谈室、水吧区、办公区、储藏室等，每个空间之间形成纵向与横向的穿插，动线分明。

办公区用简洁的装饰效果，对称的空间布局，规范方整、功能齐全，与环境形成交流，最终达成默契，突出了现代办公环境讲求效率的特征。

二楼办公区以彼埃·蒙德里安的平面构成为设计灵感，通过线型灯条对顶面进行比例分割，满足办公空间照明的同时又不失灵动性，独立办公室的设计通过调整空间建墙隔断，制造出独立的小盒子空间，嵌套入大盒子空间，使空间

形成一种"内建筑"，增加空间的趣味性，成为空间的点睛之笔。

整体空间设计运用石材、木饰面、不锈钢等结构穿插的巧妙手法呈现出门厅、办公区、水吧区的现代感与创新性。

720°旋转楼梯的设计，既作为一楼与二楼的垂直交通纽带，其流线的造型同时也为整个空间增加一定的艺术气息。

公司前台接待前厅通过木饰面与石材的几何穿插，通过木色隔栅阵列式的纵向排列、横向延展传达空间的关联性。

President Palace — China Resources
华润凯旋门

设计机构：大斌空间设计
主案设计师：金鑫

Based on the design philosophy — light is repaired, heavy adornment, golden ornaments are reasonably distributed in the whole space to naturally create a modern and luxurious atmosphere.

The wardrobes, harmonizing with the whole space in terms of color and shape, provide enough storage space for the owner. What's more, the TV wall in the living room is designed with a groove to give a sense of extension and wholeness, and the layering effect and dark wood textures also indicate the sense of fine quality.

首先项目采用轻装修重装饰的设计逻辑，将金色的点缀合理分布在整个空间内，都市现代加轻奢的感觉就很自然地被释放出来。

整体设计在满足储物需求的同时，柜体颜色与室内空间融为一体，客厅电视背景中部的凹槽设计给予了一种空间延伸感，丰富了立面的层次，同时深色的木饰面也增加了空间的品质感。

Chen
Wei

陈薇

现代设计奖 · 美学奖

2018 Top 10 Luxury Residence, Top 10 Exquisite Apartment, China Space Design Competition "Peng Ding" Awards
2018 Apartment Space Design Gold Award and Residential Building Design Excellence Award, the 6th Design Summit "Ai She Award"
2018 Annual Potential Designer, M+ China Top Interior Design Award
2018 Bronze Award, "Ginko Phoenix" Project Category, the 9th China Building Decoration Design Award
2016 Gold Award, Project Category, the 7th Red Field Prize — Interior Design Award (CIID)

曾获荣誉：
2018 年中国空间设计大赛"鹏鼎奖"十佳公寓空间，十佳豪宅空间
2018 年度第六届"艾舍奖"设计峰会"公寓空间设计"金奖，"住宅建筑设计"优秀奖
2018 年"M+ 中国高端设计大赛"年度潜力设计师
2018 年第九届中国建筑装饰设计奖"银杏金鸟"工程类铜奖
2016 年中国建筑学会室内设计分会第七届"红土奖"室内设计大赛工程类金奖

The Cry of Deer

鹿·鸣

设计机构：四川欢乐佳园装饰工程有限公司
主案设计师：陈薇

Located on the 42nd floor, this project provides a 270° view of the oversized hanging garden "From sky to ground" glass windows bring sceneries outside into the interior space while meeting the requirement for natural light.

Elements like water, wood, stone and light are used in this project. And a sense of nature is integrated into this home through the analysis of lighting angle and wind direction. At the entrance, the simple and antique waterscape comes into sight directly, and a custom curtain rushes down to surround the island where a deer herding boy stands.

In a word, the whole space is neat and refreshing.

　　本项目位于建筑体 42 层，拥有 270° 超大空中花园视野。使用"顶天立地"的玻璃窗，将楼外景观融入室内空间，同时满足室内光线及窗外景色呈现的视觉需求。

　　项目中以水、木、石、光为元素，通过对光线的照射角度和风向的分析将自然之感融入家居。在入门处，简约古朴的水景，从顶上倾泻而下的定制帘包围着承载鹿童的岛台。

木：暖意也，绵延长久，生生不息。暖木色的木材与灰色的石材交相呼应。

石：刚硬也，承载万物，长久屹立，给这个家添上了一丝坚韧与时尚。既明净，也通透自然，感官舒适，使整个空间干净，清爽。

Wood and Gray
原木与灰

设计机构：四川欢乐佳园装饰工程有限公司
主案设计师：陈薇

Adjustment to the original structure provides a space for the independent hall and an extra space for the fridge, which is more convenient for owner.

As an axis, the TV wall divides this space into two parts — one is dynamic and the other is static. The original balcony is turned into an extra space of the sitting room for family reading and interaction, which improves its functionality. Gray and white walls, as well as wood covered walls and beams, seem very simple and harmonious.

As a whole, this project is simple and moderate from color scheme to furniture and soft decoration. Moreover, the seemingly simple combination of gray walls and wooden furniture creates a natural interior space and a peaceful and comfortable feeling.

　　本项目通过对原始结构的调整，形成独立门厅空间，同时提供独立的冰箱位置，使用更加方便。

　　整个空间以电视墙为轴线，将动静区分开，原有的阳台被纳入到客厅空间的位置做留白，作为亲子互动或阅读空间，使区域功能更为多元化。色彩上采用原木、灰、白作搭配，木质上墙，包梁的处理方式让整体性更强，同时解决了结构梁外露的问题。

　　项目整体从色调，到家具形态以及软装配饰，简约舒适，不彰显不强调。浅灰的墙面，原木的家具，看似简单的搭配，为室内创造出一种自然的风范，带来宁静和舒适感。

休闲阳台

客厅

餐厅

主卧室

次卧室

次卧室

玄关

过道

主卫

衣帽间

儿童房

次卫

生活阳台

厨房

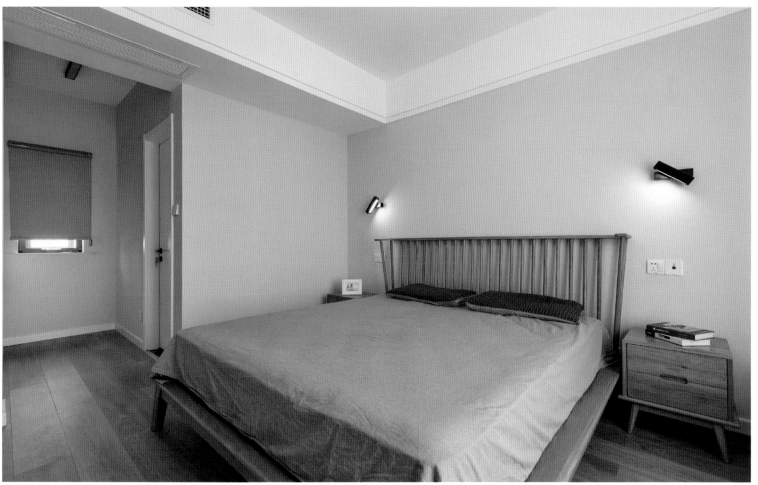

Qiao
Deshi

乔德石

现代设计奖 · 美学奖

2018 Avant-grade Designer, Annual Award of Asia Pacific Sapce Design Grand Prix

2018 Annual Potential Designer, M+ China Top Interior Design Award

2017 Most Influential Interior Designer of Liaoning

曾获荣誉：
2018 年获评亚太国际空间设计新锐设计师
2018 年获评 M+ 中国高端室内设计大赛年度潜力设计师
2017 年获评辽宁最具影响力空间设计师

Ethereality

栩

设计机构：SSD 设计事务所
主案设计师：乔德石
施工单位：SSD 设计事务所
地点：辽宁省沈阳市
面积：345 m²
主要材料：欧文莱现代瓷砖、厘芈现代家居定制、泰赫朗庭地板、莫洛尼壁炉

In this case, functional zones are divided through furniture, suspended ceiling, flooring, exhibits and lighting.

Main color scheme in this space is gray. Custom modern wood furnishing and delicate Overland tiles are all adopted to create a harmonious and moderate space.

All in all, it is designed based on the concept: from space to space itself, from material to material itself!

简约是一种生活方式，本案设计是以生活的温度为线索。

空间组织不再是以房间组合为主，空间的划分也不再局限于硬质墙体，而是更注重会客、休闲、学习、睡眠等功能空间的逻辑关系，以发散性的思维构想出各种场域情境。通过家具、吊顶、地面材料、陈列品甚至光线的变化来表达不同功能空间的划分，而且这种划分又随着不同的时间段表现出灵活性、兼容性和流动性。

方案整体以自然灰为设计的主色调，而在气质的营造上，却以生活温度为格调。现代的木色家居定制和欧文莱现代瓷砖独特的质感与细腻在整个空间内协调又不张扬；陈设饰品的选择与融合伴生出一种"内里洁净，宁谧相间"清雅的气息。

将空间回归空间本身，材料回归材料的本身！寻求一种平衡的状态，将设计回归本质之美，让生活栩栩如生。

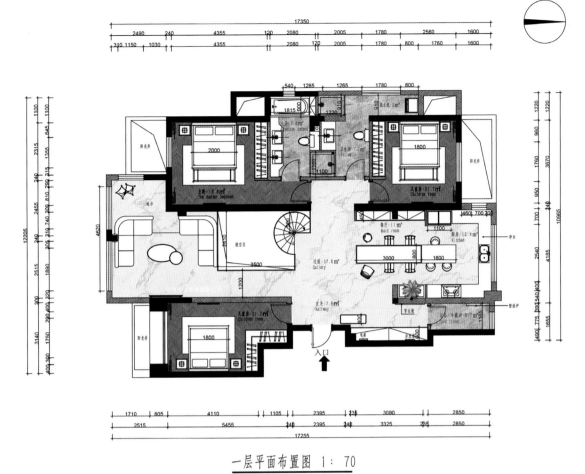

一层平面布置图 1：70

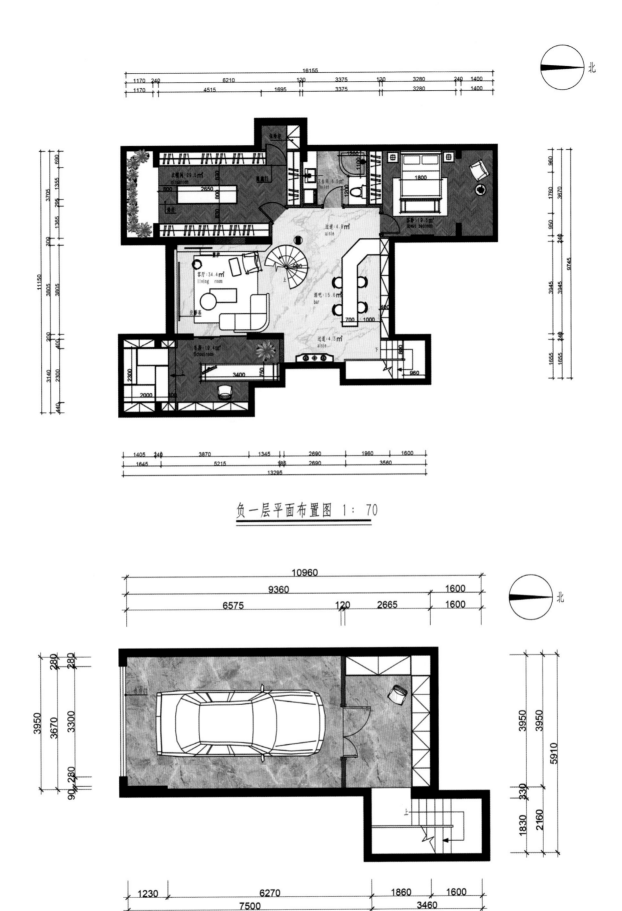

负一层平面布置图 1: 70

地下车库平面布置图 1: 70

Jin
Genchai

金艮钗

现代设计奖·美学奖

2018 Excellence Award, the 8th Nest Award
2017 Outstanding Designer of the Year, Shanghai Botao
Decoration Group Co., Ltd.
2016 Nomination Award, Shanghai Golden Bund Design Award
2015 Outstanding Designer of the Year, Shanghai Botao
Decoration Group Co., Ltd.
2014 First Prize, Shanghai Botao Decoration Design Competition

曾获荣誉：
2018 第八届筑巢奖优秀奖
2017 年上海波涛装饰集团全年年度优秀设计师
2016 上海金外滩设计大奖提名奖
2015 年上海波涛装饰集团全国年度优秀设计师
2014 年度上海波涛装饰（波澜杯）全国设计大赛一等奖

Plants Whisper

植语

设计机构：乐清波涛装饰
主案设计师：金艮钗
施工单位：乐清波涛装饰
地点：浙江省乐清市（公园一号）
面积：210 m²
主要材料：欧文莱瓷砖、梦天木门、联邦高登全屋、博洛尼橱柜、书香门第地板、杜拉维特和汉斯格雅卫浴、万物家具
摄影：创乙人·蒋居正

This project is of distinctive Neo-Chinese style that contains both classical and modern oriental charm. It aims to meet aesthetic judgment and taste of modern people and carry forward Chinese classical culture. Moreover, the perfect application and combination of the above two aspects in this space can improve its style and atmosphere immediately, which also reveals the elegance of eastern life aesthetics. In a word, this is a enjoyable, tranquil and meaningful space showing an elegant attitude towards life.

平面布置图
SCALE 1:70

注：如发现图纸尺寸与现场尺寸不符，请及时联系设计师13666609065 金良叙！

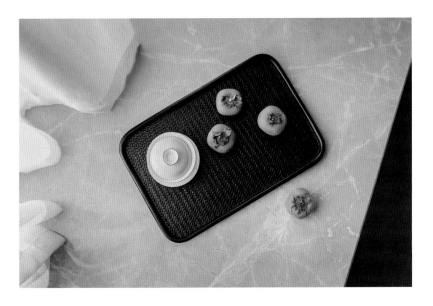

当下社会，紧凑的生活节奏已经让人们忙碌不堪，静谧的生活环境在这个物质横飞的世界里更显难寻，业主希望寻求一方生活净土，能在工作之余、生活之中享受一番雅致。

本案以业主诉求为寄托，以新中式为设计风格。新中式是兼具古典与现代气质的东方情调，以现代人的审美和眼光，来弘扬沉淀千年的中国古典文化，将两者完美融合，不但可以瞬间提升空间格调，还可以演绎出精致的东方生活美学。既有源源不息的写意，又有极致优雅的生活态度，宁静而隽永。

Zhou
Jun

周军

现代设计奖·美学奖

2017 Best 100 Villa Design Award
2017 Jury Special Prize, Red Top Award
2016–2017 Received Best 100 Boutique Home Award Twice

曾获荣誉：
2017 年的 best100 最佳别墅设计奖
2017 年度红鼎奖评委会特别大奖
2016-2017 连续二年的 best100 精品家居奖

Villa — Salute to Tadao Ando
向安藤忠雄致敬的别墅

设计机构：上海观介室内设计有限公司
主案设计师：周军

Our designer simplifies this space to lights, walls and blocks. So the wall, ceiling and floor are all white, and lines and shapes are very concise. At the entrance, a large white spiral staircase in a smooth shape extends to the roof, which not only enlarges the space, but also becomes a medium between two walls to create a layering space. There are no redundant ornaments except some necessary Bauhaus furniture.

　　"空"，是它给我的第一感受。但这种"空"并非"缺失感"，运用最简单的构成原理，将空间简化至光、墙、体，从而使本案空间内容成为被关注的主体而包罗万象，如一股能量气场将人怀抱。高敞的空间中，墙面、天花、地板几乎完全留白，线条、轮廓极致简练；进门处大体量的白色旋转楼梯造型流畅，向上延伸的姿态既增添了建筑的尺度感，亦扮演起面与面之间的媒介物，交织出极富层次的空间美感；空间中几乎无任何多余的装饰与物件，包豪斯式的必要家具零星点染其中，形成一场"有"与"无"间的对话，在对话中，每个角色都有了自身的内涵与光芒。

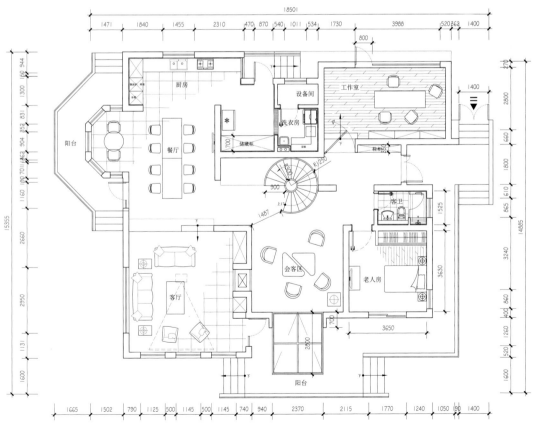

一层平面布置图 1:80

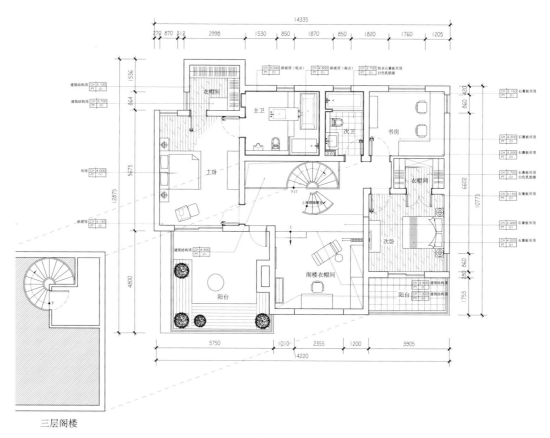

三层阁楼

二层平面布置图 1:80

Shi
Weijun

施伟军

现代设计奖·美学奖

2016 Elite Designer of Asian-Pacific Region, Mango
Prize—Be Your Best
2016 China Decoration Design Award, CBDA Design
Award and CBDA Special Contribution Award
2014 Most Influential Designer of China, National Top
100 Versatile Designer Award
2013-2014 Outstanding Designer of Taizhou Region,
Asia Pacific Interior Design Awards for Elite
2013 Excellence Award, the 4th China International
Space Environment Art Design Competition (Nest Award)

曾获荣誉：
2016 年度 / 王的盛宴 /【芒果奖】亚太地区精英设计师
2016 年中国装饰设计奖 /CBDA 设计奖 /CBDA 特别贡献奖
2014 年全国百佳设计综合典范人物奖 / 全国最具影响力设计师
2013-2014 亚太室内精英邀请赛 / 台州分赛区 / 杰出设计师
2013 年第四届中国国际空间环境艺术设计大赛（筑巢奖）中优秀奖

Jiangnan First Mansion
江南首府

设计机构：风语筑·展览·装饰·设计
主案设计师：施伟军

A combination of artistry and modernity — feature of flat villa.
A conversation between nature and man-made spaces;
A sequential arrangement of the ceiling and wall tiles and openness of space;
A blur zone between spaces...

Outside, a world of nature,
Inside, a world for free growing;
The orderly space inside contains the humility of a traditional Chinese
garden view,
Harmonization,
So the invisible life in this space goes on...

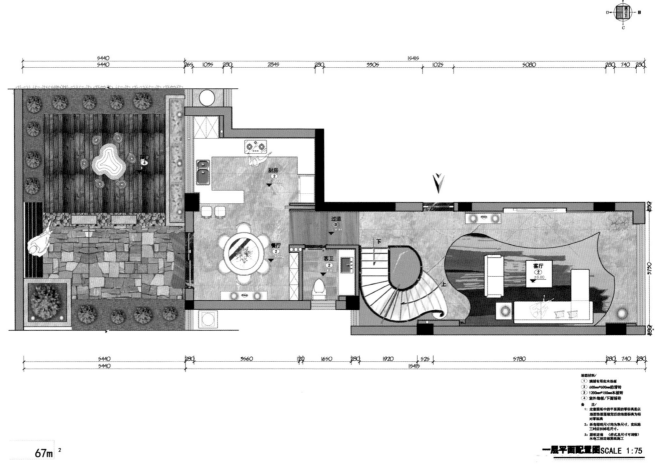

67m²

一层平面配置图SCALE 1:75

现代与艺术的结合 —— 打造平墅特色。

环境与人造空间对话的延续；
天花 / 墙面砖的理性排列与户外的开放；
环境间有着模糊的地带……

外在是自然的天下，
无拘无束地生长着；
内在则以秩序却含着中国园林中宇宙观的谦卑，
互相搭配，
使得"场域"那不可视的生命被延续了……

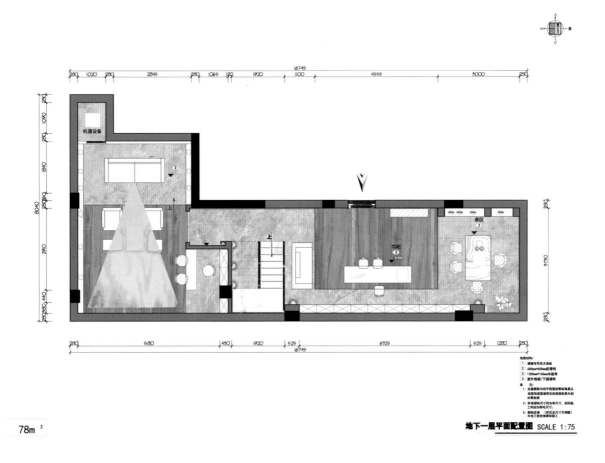

78m²

地下一层平面配置图 SCALE 1:75

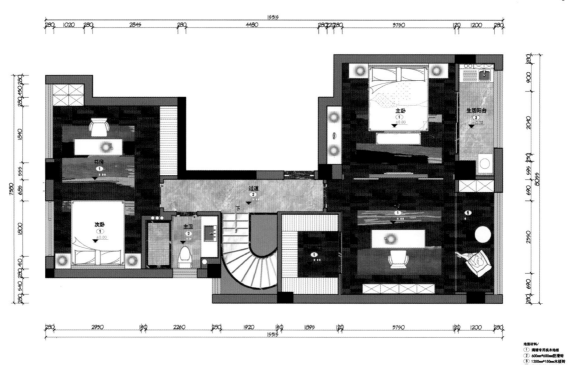

88m²

二层平面配置图 SCALE 1:75

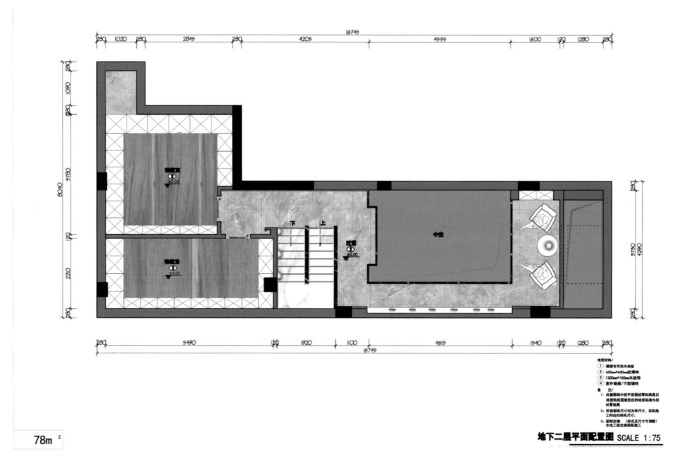

78m²

地下二层平面配置图 SCALE 1:75

Central Mountain Residence
中央山公馆

设计机构：风语筑·展览·装饰·设计
主案设计师：施伟军

Chinese traditional design is generating more and more new forms and ideas on its own path.

对于中国的设计来说，新的形式、新的创作道路正是要从中国自己的传统设计格局里产生。

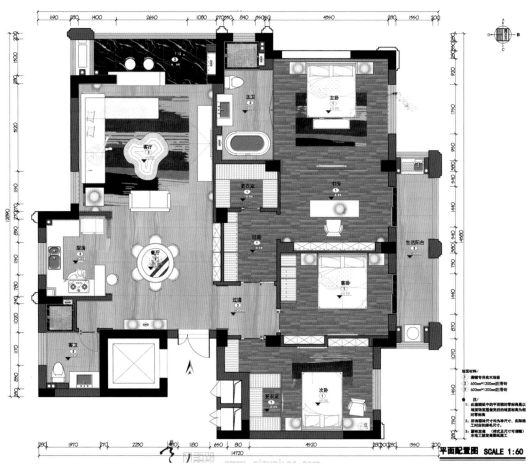

158m²

平面配置图 SCALE 1:60

Modern Design Excellence Award
现代设计奖·优秀奖

Huo
Cheng

霍成

现代设计奖·优秀奖

2018 Top Ten Designer, the 2rd OPPEIN Dream
Design Home
2017 Top Ten Interior Designer of Suining
2016–2017 Champion of the Year, Yenova
Decoration Suining Department

曾获荣誉：
2018 第二届欧派杯梦想设计家十佳设计师
2017 遂宁十大优秀室内设计师
2016-2017 业之峰装饰遂宁分公司年度冠军

Shanyu Lake

山屿湖

项目设计：霍成
地点：遂宁 山屿湖
面积：228 m²
风格定位：现代风格
摄影：霍成
主要材料：瓷砖、墙纸、硬包、石材、乳胶漆

The construction area of this project is 228 m², and the original 4 rooms are turned into 3 rooms. With careful consideration of lines, faces, proportion and materials, the designer creates a spacious space with top quality. What's more, the original staircase is removed because it was too narrow and useless, and now here is redesigned into a display space for the sitting room.

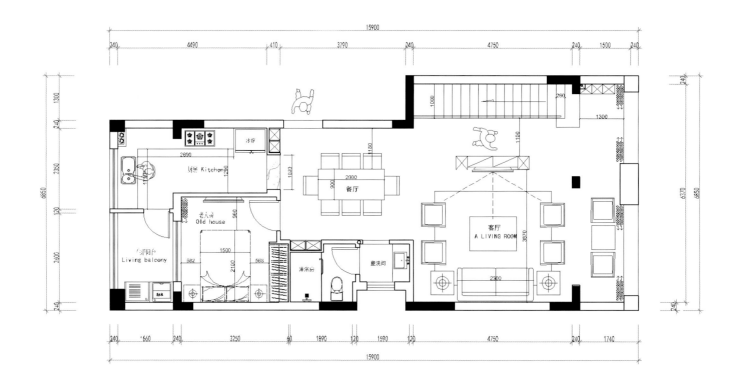

本案建筑面积 228 m²，原有的 4 个房间改成了 3 个房间，通过线、面、比例以及材质的运用，营造极强的空间感和质感。原有的楼梯空间较窄没有太大的实用意义，将之拆除后重新设计，将客厅空间延展至此，成为客厅的展示空间。

男业主是 90 后生意人，本案中多处运用黑色材质，这些细节就如男人气质中的冷静、硬朗。

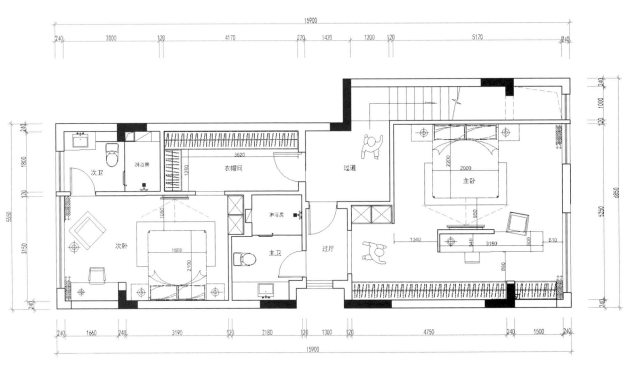

Chen
Yunjian

陈运健

现代设计奖·优秀奖

Minimalism
极简主义

设计机构：吾蕴空间
主案设计师：陈运健
地点：湖州
面积：400 m²
风格：极简风格
主要材料：水泥砖、水曲柳、乳胶漆

As for this case, the main color scheme is warm white, cool gray and deep blue, and the latter two can be found on the floor and the curtain respectively.

Faces and lines are well organized and combined through spatial design techniques, such as extending the panels vertically and edging them linearly.

The open living room and dining area is featured by a very modern fireplace. And an open kitchen is right beside the dining area to enhance interaction. The master suite is equipped with a private bathroom, a study room and a big cloakroom, which can reveal the owner's requirement for a high quality life. Carefully selected furniture and delicate ornaments are well arranged to present an elegant and artistic style.

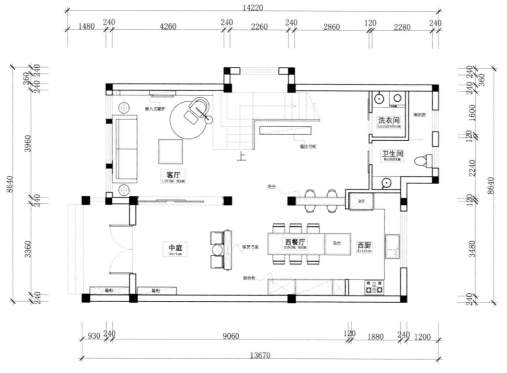

14220

1480 240 4260 240 2260 240 2860 120 2280 240

240 240 360

360 240 240

3960

8640

240

3360

240

嵌入式壁炉

壁挂书柜

客厅
LIVING ROOM

上

洗衣间
Laundry room

淋浴房

卫生间
WASHROOM

120 1600

120 2240

120

中庭
Atrium

铁艺书架

吧台

餐具柜

西餐厅
DINING ROOM

吧台

ICE

西厨
Kitchen

3480

930 240 9060 120 1880 240 1200

13670

一层平面布置图 1:75
注:具体应以施工现场放线尺寸为准

本案设计空间主体色调被赋予阳光的暖白，结合客、餐厅中地面的灰与其后那一抹高雅的湖水墨蓝色垂帘，将"冬"的缄默冷静勾勒其中，赋予空间不同的情感⋯⋯

纵向面延展、线性收边的空间设计手法，将创造面与线的完美结合，丰富了艺术形态在简单空间里的多层次变化。

当体验者身处客、餐厅区域时，开放式空间顿时带给体验者的是空间的开阔与震撼，而与现代感十足的壁炉完美结合，更是为冬季主题增加了温暖与柔和。

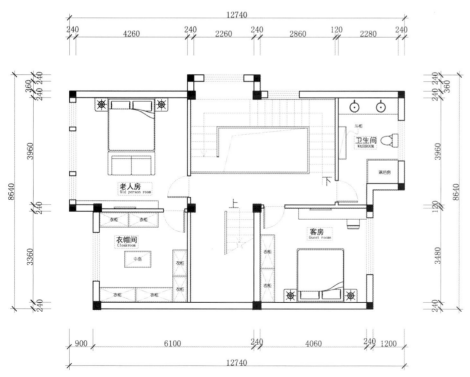

二层平面布置图 1:75
注:具体应以施工现场放线尺寸为准

舞动的火苗，温暖了整个冬季。紧邻餐厅区域的是开放式厨房，打破了原来的封闭状态，与餐厅、客厅相连接，使居家空间更加通透宽敞，增强互动性。在餐厅的吧台处，享受甜蜜的下午茶或沉醉在鸡尾酒的微醺下，配以舒适、温暖的场景，给人一种对未来美好生活的向往。

在主卧套房内打造具有独立卫生间、书房及强大衣帽收纳系统的整体功能，体现出生活的高品质需求。整体空间配以考究的家具、精致的饰品，并合理摆放，呈现出为注重生活品质和格调的精英人士所定制的居家环境。

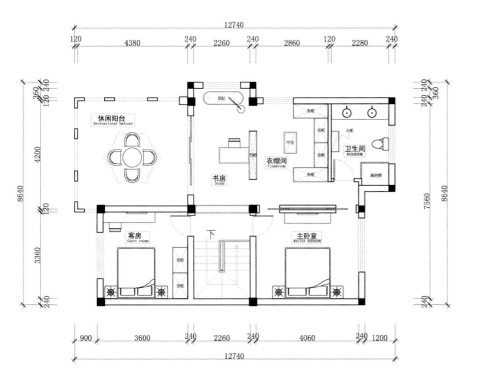

休闲阳台
Recreational balcony

书房
Study

衣帽间
Cloakroom

卫生间
WASHROOM

客房
Guest rooms

主卧室
MASTER BEDROOM

三层平面布置图1:75
注:具体应以施工现场放线尺寸为准

Peng
Hong

彭鸿

现代设计奖·优秀奖

Interior Design Certificate, China Interior Design Association

曾获荣誉：
中国室内装饰协会室内设计师资格证

Moyi Design Club

墨亦设计会所

设计机构：墨亦设计
主案设计师：彭鸿

In this case, the optimum combination of Nordic and Japanese style, the proper division of static and dynamic spaces, and the right decoration of green plants make the space cozy, cool and natural. The concept of Nordic and Japanese style is "less is more", so the space is given plain hues to produce austere visual effect. Working in such a natural and comfortable space, people can feel calm inside.

　　北欧与日式风格的完美结合，空间动静的完美划分给人温馨清爽的感觉，室内绿色植物的点缀传达出自然的气息。

　　北欧和日式风格推崇的就是"少即是多"，素色的基本调本身自带极简的视觉效果，这样的办公空间安神定心，自然朴素又不失温馨的氛围。

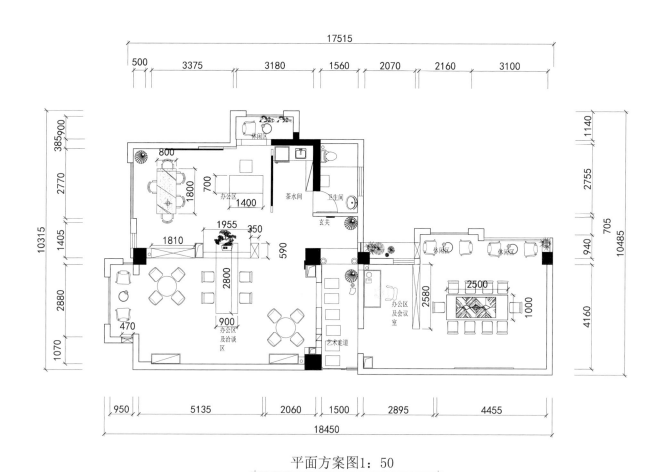

17515

500 3375 3180 1560 2070 2160 3100

385900
2770
10315 1405
2880
1070

800

1800

700

办公区
1400

茶水间

卫生间

玄关

休闲区

1955 350

1810

2800

590

470

900
办公区
及洽谈
区

艺术走道

休闲区 休闲区

办公区
及会议
室

2580

2500

1000

1140
2755
705
940
10485
4160

950 5135 2060 1500 2895 4455

18450

平面方案图1：50

Chen
Jiarui

陈嘉睿

现代设计奖·优秀奖

2017 Elected as Life Aesthetician of the Year by Asia Pacific
Designers Federation
2017 Best Apartment Interior Design Award of the Year (Project
Category), the 3rd Circlerider Awards
2016–2017 Asia Pacific Designer
2016–2017 Silver Award in Show Flat Category, Red Top Award
2016 Top Ten Residential Designer of Hubei

曾获荣誉：
2017 亚太空间设计年度评选生活美学家
2017 第三届师客莱德奖作品类"年度最佳公寓
空间设计奖（工程类）"
2016-2017 亚太空间设计师
2016-2017 年度国际环艺创新设计作品大奖赛
（华鼎奖）样板间空间类银奖
2016 年湖北年度十佳住宅空间设计师

Office of Line Fine International Finance

联发国际金融办公室

设计机构：一筑设计事务所
主案设计师：陈嘉睿
地点：中国武汉联发国际大厦
面积：238 m²

This case is a combination of classic black and white and blue hues, which breaks people's rigid cognition of commercial space. Light and shadow, soft decoration and green plants form a flowing picture with delicate details and perfect proportion in the space.

近似没有温度的外表下其实隐藏着炽热的心，"隔离是创造力的天敌"。我们通过平面布局创造一个具有意识而不是形态上的空间来促进员工在共享空间里融洽舒适的互动。仅仅是利用一面展示柜，用格栅立面造型来拉伸空间高度，底部订做成品仿真壁炉，通过这样一个整体的背景造型来引发访者对流动空间动感的想象。当你再进入这个空间时，会议室玻璃"墙体"的底部周边设计了半隐藏式的灯光，仿似在神秘的黑色空间中乍破的天光倾泻而出。这一切正如托尔斯泰所说："幻想里有优于现实的一面，现实里也有优于幻想的一面。完美的幸福将是前者和后者的合一"。

整个办公空间，我选择用"完美灰色"演绎，灰色能够在视觉上带来极好的和谐感，给人脑的信息处理压力低，让人内心更平静、祥和；反之很多颜色过于刺激，会制造处理压力，让人有压抑感和潜在的恐慌感。高雅装饰更是让这种境界升华，搭配冷暖色调更是相得益彰。高级灰可以去除繁杂，让空间极具穿透力，有种与身俱来的魅力。

室内地面使用了灰色大理石交错铺贴，更细腻地表现了纯粹高洁的环境基调，搭配着白色的天花，保证了宛若天光的均衡照明。大面积的落地玻璃，消弭办公室内外的边界，可纵观长江二桥景色。

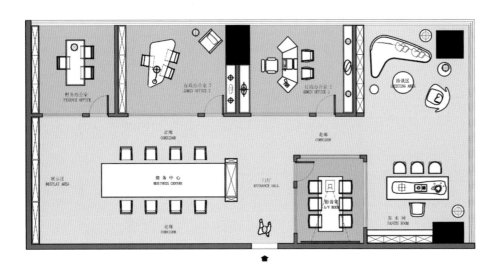

"最好的建筑是这样的，我们居住其中，却感觉不到自然在哪里终了，艺术从哪里开始。"林语堂的这段话，便是我们的设计灵感之初。像大部分企业办公空间一样，有些必需的功能空间与配套，从办公、会议、到休闲、餐饮……但我们真正要做的除了功能性的要求，对于员工而言，工作环境也必需要有足够吸引力，可以为其带来独特的工作体验，大胆、个性、创意十足的工作环境是能使员工快乐地工作的，而自由的办公空间更是能产生无限的可能。

会议室同样选用透明玻璃作为"墙壁"，隔绝出相对隐私的私密空间，其余时间则恢复透明，维持整体空间的流畅性。主色调是还是以低饱和度的高级黑为主，视觉上更趋于沉稳、安定的会议氛围，配以黑色几何型坐椅。

光影、软装配饰、绿植伴生着流动的空间画面，在不同的动线之下，皆能步移景异。也正是因为这些含蓄的细节，完美的比例，包含在整体空间之内，才能有了这润物细无声的效果。

　　办公空间不应该只是一个赤裸的商业型场合，理应承载除商业价值以外更多美好的东西，可以有趣、个性、简单而具有企业特有气质的综合性办公空间。在逐渐深入的过程中你会发现，这必然不只是一个关于空间的设计，而是关于如何把企业文化转化成工作方式、生活方式的一次探索。无论是经典的黑白灰，还是锦上添花的蓝色格调，我们的设计态度，就是要让美这件事发生在已经被固化的刻板商业空间的认知上，而不是继续框住这种认知，从而妨碍设计。正如 Philippe·Starck（飞利浦·斯塔克）所说"设计是拒绝任何规则与典范，本质就是不断地超越与探索。"

Xu
Hongyu

徐宏宇

现代设计奖·优秀奖

2018 The 13th Huading Award
2017–2018 Top 100 New Designers of China Interior Design
2017 Silver Award of the Diamond Cup of China Interior
Design Award
2017 IDEA-TOPS
2016 Excellence Award, International Art Design Award (IADA)

曾获荣誉：
2018 第十三届华鼎奖
2017–2018 年度中国室内设计百强新锐设计师
2017 中国室内设计大赛金钻杯银奖
2017 艾特奖
2016 (IADA) 国际艺术设计大赛优秀奖

Super General Korean Restaurant, GIDEAR

将军牛排韩国餐厅金安国际店

设计公司：杰夫设计
主案设计：徐宏宇
地点：哈尔滨

Due to the particularity of this space, there is no shed on the top floor of the shopping mall, which, on the contrary, can bring natural light into the space. Hence this space can be redefined by nature and new conception. Stepping into here, one can directly see an art installation — Flos Hibisci, also known as "desert rose". As the national flower of Korea and an artistic symbol, it is not only full of natural beauty, but also gives out a sense of modernity. The antique handcraft sheepskin lamp forms a sharp contrast with the traditional Korean wardrobe, which blends traditional Korean culture with artistic elements in the whole space.

　　光让大自然焕发了生命力。本案由于空间特殊性，在商场顶楼，没有棚，白天太阳光照充足，恰好可以借自然结合理念重新定义空间。进入餐厅后，首先看见的是绽放自然美感的艺术装置——木槿花，韩国国花木槿花又名"沙漠玫瑰"，木槿花装置贯穿大厅，形成现代感极强的艺术符号。古朴的手工艺羊皮灯和充满历史年代感的韩国传统柜子形成了鲜明的对比，将韩国传统文化与整个空间里的艺术元素糅合在一起。设计师希望用这种表现形式让空间焕发出新的生命力。

He
Dongqing

何冬青

现代设计奖·优秀奖

2017 Study Tour in Singapore
2016 Study Tour in Copenhagen, Denmark
2015 Received Further Education in DOMUS College, Milan, Italy
2013-2014 Excellence Award of Nest Award
2013 Excellence Award of Guangzhou Design Week

曾获荣誉：
2017 年到新加坡游学
2016 年到丹麦哥本哈根游学深造
2015 年到意大利米兰多莫斯设计学院深造
2013-2014 年筑巢奖优秀奖
2013 年广州设计周大赛优秀奖

Jinyu Yuefu Modern House

金隅乐府洋房

设计机构：唐山丁乙空间设计有限公司
主案设计师：何冬青
地点：唐山市

This project has two units with an elevator. By creating abstract structural forms, our designer uses new materials, modern crafts and lighting effect to enrich this space. What's more, various mirrors, stainless steels, polished granites and marbles are also adopted to decorate the walls. As for lighting, new lighting installations project and reflect on the metallic and mirror materials to create shining and gorgeous effect, which makes the modern materials and crafts more attractive in this concise and lively space.

平面布置图

　　该项目位于河北省唐山市北新东道与龙泽南路交叉口东 400 m 处，与大城山公园仅有一路之隔，属于一梯两户电梯直接入户设计，也是低密坡地洋房景观社区。周边休闲、文化、医疗、金融、商业以及教育设施齐全，统一供水供电供暖，承接大城山绿荫肌理，与凤凰山公园相辅相成，搭配以经典的装饰艺术建筑风格，建筑立面是在精炼、标准的古典形式上所进行的改良，所有细节得体。

　　在本案中设计师利用了抽象形体的构成，展现新型材料及现代加工工艺的精密细致及光亮效果，在室内大量采用镜面及不锈钢、磨光的花岗石和大理石等作为装饰面材。在室内照明上，采用投射、折射等各类新型光源和灯具，在金属和镜面材料的烘托下，形成光彩照人、绚丽夺目的效果。并在简洁明快的空间中展示了现代材料和现代加工技术的高精度，传递着时代精神。

Yang
Xiaotian

杨啸天

现代设计奖 · 优秀奖

2017 Bronze Prize, China Soft Furnishing Art Forum and the
First CBDA China Soft Furnishings Arts Festival
2013 First Prize of the 3rd International Innovative
Environment Art Design Competition
2011–2012 Most Influential Designer Award (Dining and
Entertainment Space)

曾获荣誉：
2017 年中国软装陈设艺术高峰论坛暨首届中国建筑装饰协会软
装陈设艺术设计节铜奖
2013 年度第三届国际环艺创新设计大赛一等奖
2011–2012 年度最具影响力设计师（餐饮娱乐空间类）奖项

Shijiazhuang Jufulou Restaurant

石家庄聚福楼餐厅

设计机构：石家庄啸天装饰设计有限公司
主案设计师：杨啸天
面积：670 m²
地点：石家庄

In this case, based on neo-Chinese style, the designer makes full use of the texture of wood to convey a concept of nature and health, which exactly fits the feature of traditional cuisine — Jiaozi in this Restaurant. Lights, wall painting, mountain and water landscape are creatively combined together to convey the artistic beauty of a space, which is also a perfect fusion of Chinese cuisine and design conception. Therefore, this spectacular and harmonious catering space can also spread the auspiciousness and filial piety of Chinese food culture through modern and traditional elements.

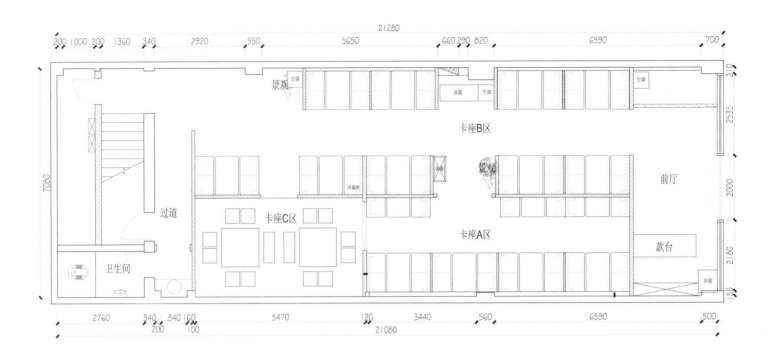

景观　空调
冰箱　空调
空调
卡座B区
冷藏柜
景观
前厅
过道
卡座C区
卡座A区
款台
卫生间
冰箱

21280
200 1000 200 1360 340 2920 550 5650 660 290 820 6590 700
210
2535
7080
2000
2180
155

2760 340 840 160 5470 120 3440 560 6590 500
200 100 21080

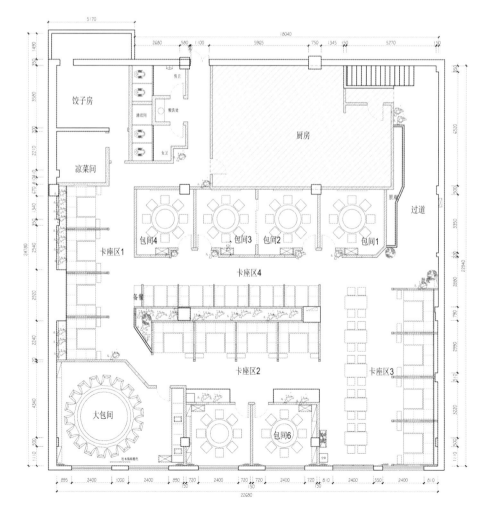

孝在餐桌，味在人间，缘聚聚福，乐享亲情，设计师将中华美食的精髓与设计理念完美融合，传播了聚福楼现代与传统和谐并存的中国福孝餐饮文化。

本案设计立足于新中式风格，采用原木材质，给人一种天然健康的概念，很是符合本餐厅传统美食饺子的特色。利用山水景观、灯饰、墙绘等营造意境美，创造了与众不同的餐饮空间。

Yang
Huansheng

Guo
Shihao

杨焕生　　郭士豪

现代设计奖·优秀奖

Yang Huansheng established YHS Design Agency in 2005. And in 2007, its team was entrusted to design the Mulan Boutique Hotel in Taipei, which is a very successful project for YHS. And from that time, YHS has become a popular and professional commercial space and hotel design team.

YHS stands for Yearning for the Highest Satisfaction, which means that the ultimate goal of our team is to meet our clients' requirements in every case. And YHS can also stand for Youthful, Holistic and Stimulating.

杨焕生设计师 2005 年创立了同名事务所。2007 年，YHS 设计事业受托打造台北沐兰时尚精品旅馆，凭着成功优异的设计成果，YHS 设计事业一战成名，成为业界最炙手可热的商业空间及旅馆设计专业团队。

YHS 设计事业意义是 Yearing for the Highest Satisfaction（渴求最高的满意度），意味着我们无论接受任何案件，终极目标都是追求团队及业主双方最极致的满意度，而这三个字母也可以是 Youthful（ 朝气活力 ）、Holistic（ 整体全面 ）、Stimulating（ 激发感动 ）。

我们的设计创作，力求摆脱陈腐、与时俱进，所以即便是向经典致敬也充满了当代活力，在设计的范畴上，我们关注的不仅只是单一空间，更会从空间、基地与环境三者的连结来考虑，让设计的氛围与功能性更加整体全面，另外，我们擅长解构既有材质，重新赋予结构重组的新生命，使设计带来视觉上的全新冲击与感受，我们在细节处理上亦是如此，希望能打动使用者。

Azure Space

天青

设计机构：YHS DESIGN 设计事业
主案设计师：杨焕生、郭士豪
面积：211 m²
主要材料：木皮、钛金属、订制家具、花艺、订制灯、镜面、大理石

This case creatively combines modern life with oriental elegance, so as to integrate life inside with urban environment.

An artistic charm lasts in the whole space. In the living room, a wall of metal lotus turns into a picture, and the sliding screen can make it an open or private space. Dark wood decorations, neat lines and curved frames are alternatively appeared to create a false impression that time stands still in the space. And there is a sharp contrast between the living room and dining room.

In the well-designed corridor with elaborate visual points, the vision is narrowed, which makes ones feel the generous, elegant, luxurious and exquisite compositions.

The dynamic lines in the space are carefully arranged through symmetry, falseness and trueness , view-borrowing and other techniques to bring natural scenery into the space. Hence, this refreshing space is dominated by green plants, water and flowers which allow the owner to enjoy the fun of movement and quietude.

引景造境动静皆趣

本案设计将美学品味与空间美感体现于时间动态之中，希望藉由设计统一空间调性，体现文化底蕴与当代空间之美，结合创新视角将现代生活与东方雅韵相融合，使居住生活与都市环境相融合。

位于客厅的金属画作特意模拟荷花之美，藉由屏风虚实交迭开合，隐蔽于客厅深处，使优雅艺术气氛蔓延于室内，深色木作与工整线条搭配弧形收边作为表现，藉由静瑟氛围创造时间停顿之感，客厅餐厅的开阔舒适形成强烈对比。

廊道特意采用限缩视觉景深并且精心装饰，创造出精致的视觉亮点，行走其中让人感受到大气、清雅、华丽、精巧的构图变化。

本案在空间设计中精心铺陈动线韵律及视景，采用大量的对称、虚实、借景等手法，使环境景色透入室内，映然生意，空间中漂散绿意、水气、花香；居住者可以细细品味动静皆景的趣味。

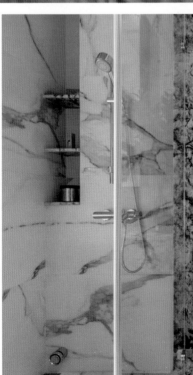

Metallic Space in Woodland

森铸

设计机构：YHS DESIGN 设计事业
主案设计师：杨焕生、郭士豪
面积：240 m²
主要材料：木皮、钛金属、原装家具、订制灯、镜面、大理石

In this long and deep space, the designer uses a horizontal axis to maximize the owner's view of natural scenery. And a metallic sliding door reorganizes the space to deepen the view. Most importantly, outdoor sceneries are borrowed to enrich every corner.

As for space design, lighting and texture are two woopdant aspects. Natural light comes from each big window, and the changes of light and shade in a whole day can be observed. A metallic door is featured with subtle hair-like patterns; gray marbled and wood texture transfer visual enjoyment to tactual enjoyment.

The cambered ceiling shows its attraction and elegance through a curved line, which reflects designer's concerns for spaciousness, dynamic lines, measurement, relation, function and aesthetics.

本项目在离离蔚蔚一片茂盛的森林公园边，择一处浓绿气息怀夹人文荟萃雅质苑所。

"居"，是一种生活容器也是记忆载体，透过设计以长时间酝酿及细致工法堆栈，来期望将来能留存美好生活轨迹。

长景深的空间里，用创造水平轴线的延续来衔接窗边天地大景，以金属细致横向活动拉门重新介定区划空间关系深化视野旷貌，将屋外绿意大大方方借景堆栈于室内每一角落，达到空间中处处揽景无遗。

光线及材质是本项目空间设计的主角，大片通透的采光窗让一天 24 小时光影都具清晰变化；发丝般细致纹路金属门搭配灰色大理石及温润木皮光泽，将空间细节由视觉提升为触觉，亲触里荡漾的美好的回忆，使人置身其中共酿沉着韵味，心境自然和缓沉淀。

曲形天花的设计将空间、动线、尺度、关系、机能与美学在一道连续弧形线条中描绘道尽同时也召唤出一份居所具备的性感与优雅。

Lin
Zhengwei

林政纬

现代设计奖·优秀奖

2019 Awarded TID Award, Inside Award shortlisted
2016–2018 Awarded Germany Red Dot Design Award, iF
Award, Golden Pin Award, A'design Award Italy and K-Design
Award Korea in commercial and residence field
2013–2015 Awarded China IAI, JINTANG PRIZE, Golden
Pin Award
2011–2012 Awarded TID Award 2011 / Heritage – Architecture
2011 Co-lecturer with Japan architect, Keisuke Toyoda, in
Grad Inst of Architecture NCTU
2010 In charge of Design Director/ Snuper Design Inc.

2019 荣获台湾室内设计大奖，Inside 世界设计大奖入围
2016-2018 荣获德国红点设计大奖、iF 设计大赛奖、金点设计奖、
意大利 A' 设计大奖、韩国 K-Design 设计大奖（商业住宅空间）
2013-2015 荣获中国 IAI 亚太设计联盟设计奖、金堂奖、金外滩奖、
台湾金点设计奖（住宅空间）
2011-2012 荣获台湾室内设计大奖
2011 与日本建筑师 Keisuke Toyoda（丰田启介）在交大建筑研
究所共同授课
2010 成为大雄设计事务所总设计师

Malaysia Mansion
印象马来西亚

设计机构： SNUPER DESIGN / 大雄设计
主案设计师： 林政纬
地点： 马来西亚
面积： 334 m²
施工： inD'finity Design (m) Sdn. Bhd.

After discussing with the owner, our designer decides to rebuild a natural and humanistic space with careful attention to material, layout, details and Malaysian multiculturalism.

The TV wall is extended, which becomes a complete and generous panel for the porch. The large window on both sides brings natural breeze and light into the space. This space is also a fusion of culture and nature, which can be felt from the natural elements in this space.

时间像洪流，随着它的脚步带走过往的事物，也留下了生活的足迹。设计，真实反映了文化、历史和人的价值，如何将大马四季如夏的地域性特征，建构至百坪独栋透天厝中，是本案设计师的首要课题。

独栋别墅有着四周庭园环绕的优势，空间策略也因此涵构特色而有了呼应。

设计师和业主沟通后，从材质选用，空间切割至细部美学，结合大马地区的多元文化特性，从设计面重塑自然与人文的共构空间。

阳光、空气、微风，温暖的、潮湿的，这些浸润皮肤的生活气味，彷佛是热带岛屿特有的无形记忆。透过开阔、方正的格局，将室外的原生绿意与室内的公领域相互串连，流通的动线规划引介内外空气流动，消弭了场域的分界。

In the open living room on the other side of the axis, windows are decorated with small horizontal panels which are so austere that the pavilion and plants outside become the spotlight. Floor here is of different height, and spaces are distinguished by different materials. Italy REX tiles — French golden tile series are adopted to decorate the living room. Other different materials with natural texture are also applied to the space, such as Angola pearl, galaxy grey and pine pattern tiles, to interpret the beauty of movement and quietude.

Granitic TV wall and wood texture grille extend to the ceiling. Ordered veneer and lighting not only highlight the grand spatial layout, but also stretch the walls.

The open kitchen is accessible from here. And the unique cellar symbolizes owner's taste of life.

As a socializing zone, the open dining room and kitchen are connected by the well-ordered veneer on the ceiling. However, the floor is covered by tiles which are cut in proper proportion. The delicate texture shows the wonderful visual effect from aesthetic aspect.

Decorated with wood texture frame, the ceiling in the hallway presents layering and warm lighting effect, which shapes a sharp contrast with the coolness of the marbled floor.

本案设计将电视主墙尺度加长，让迎宾玄关有了完整而大气的回廊切面，两侧的开窗引流室外的微风和光线，植入原粹的自然生态，将场所的人文和第二自然精神带入生活体验。

轴线的另一侧则为开阔的降板客厅，降板的低调设计，让视觉迎接了窗外的发呆亭与绿意。厅区间的地坪高低差，以材质差异作为区隔，客厅地坪选用意大利原装进口 REX 磁砖法国流金系列，安格拉珍珠、银河灰、松柏石等相异材质做自然纹理的界接，型塑动静交错之美。

电视墙粗犷的花岗石皮与交织的木纹格栅，扩展至天花的木皮格栅与镜面，以序列的木皮和灯光，展现大气的格局设计，拉大了尺度也延伸了表面空间。

转折入内则是主人的开放餐厨空间，独特的酒窖设计凸显了主人的生活品味。

在那镀上厚厚一层又一层东方的金色韵味中，矩列的木皮收纳柜，保存着味道和传承的记忆，提醒着我们生活的本质和意义。

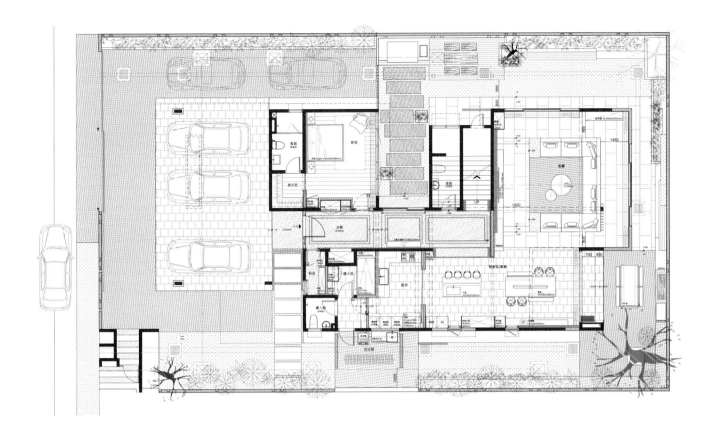

Ten kinds of tiles and stones, metal and veneer are alternately used in the entire space. Therefore, there are sharp contrasts between the roughness of stone and gentleness of wood, the coolness of marble and transparency of glass. From the vertical grille to the ceiling, the space gives a sense of wholeness, but the linear light leads our vision to different zones in the space. All in all, diversified materials add not only the feel of coolness and comfortableness but also luxuriousness and feeling of vacationing to this tropical residence.

开放的客餐厅和厨房作为生活的交谊场域，藉由天花序列木皮包梁贯穿，而地坪使用比例切割的瓷砖计划，在美学表现上，可窥见纹理之间细致的视觉效果。

廊道顶部采用木皮框构出富有层次的灯光，兼具温润、明亮的意象，与地坪清冽的大理石，形成鲜明的冷暖对比，勾勒出兼容的材质效果。

全室材质使用了十种瓷砖，与石材、金属、木皮交互搭配，在细部的美学中再现了岩石的粗犷与木质的温润，大理石的清凉与玻璃的通透，从立面的排列格栅沿展至天花，其中穿插具引导效果的线性光带，无形中定义出隐喻的场域分界。多元的材质，为热带住宅添增了舒适的清凉感，同时不失奢豪与度假风情。

椰子树清新的气味伴随着夏日的梦境。在暖黄的灯光之下，钻进凉爽的被褥中，享受一夜好梦。

Ma
Jiankai

马健凯

现代设计奖 · 优秀奖

Category: building renovation, normal layout, big space, commercial space, etc.

Style: modern simplicity, modern luxury, functional design, minimalist style, natural humanistic design

Conception: Jie Yang — black and white, fashionable and avant-garde
Da Si — natural, humanistic, elegance
based on two different concepts "Jie Yang" and "Da Si" to create spaces in different styles
Spatial design should satisfy the user's actual needs and create an amplified spatial effect with reasonable furnishing and aesthetics to present a spacious, cozy and ideal home at last.

设计类型 / 旧屋翻新、一般格局、大坪数、商业空间等等……
设计风格 / 现代简约、现代奢华、机能设计、极简风、自然人文

设计理念 /
将界阳 & 大司室内设计，打造出两种截然不同的风格，
界阳 = 黑白时尚前卫、大司 = 自然人文典雅的品牌概念。
设计以放大空间效果为前提，将空间使用者的实际需求及机能，
落实为合理化的配置及美学安排，使完工后的作品呈现更为宽敞
大气、舒适快意的理想居宅样貌。

Imperial Style
帝苑气度

设计机构：界阳＆大司室内设计有限公司 /Jie—Yang Interior Design
主案设计师：马健凯
地点：中国台湾桃园市
面积：340.5 m²
主要材料：镀钛、锈化镜、石皮、钢刷木皮、特殊漆
摄影：Yana Zhezhela, Alek Vatagin

As a residential architecture specifically designed for retired entrepreneurs, the house layout primarily considers the couple's life style. The broaden public space, divided into areas on the basis of residents' favored functionalities, brings closer the couple's intimate relationship.

The public space, on the right of living room, consists of the dining room, exercise room and multimedia area. The big swivel TV, centered on the public space, is watching-convenient ubiquitously. The master suite directly behind the open study room is on the left of living room. The sliding doors, furnished between the study and living room, are life-thoughtful for either the doors opened make the study room a part of master suite, or the doors closed divide the study room from the bedroom as a private space.

此宅居为企业家退休后的乐活宅邸，起居动线以夫妻两人为主。设计师将横向通透大公领域空间在居者熟悉的生活机能下作区块划分，让两人在透过生活动静间，加深彼此的互动。

空间以大厅区域为中界，右以餐区、吧台区、运动区、视听区串起的公共区域为主，并于其中巧妙配置可旋转大屏幕，任何角度皆可供观赏娱乐。左侧主要以开放式书房后衔接的大主卧空间为主。在书房与客厅分界中装置有大面推拉隔门，当隔门展开时书房即为主卧的一部分，收起隔门时书房则界定为独立区域。

进入宅邸，映入眼帘的是迎宾的大气势，在玄关的对接中展开序幕，带来

犹如帝苑级的格局礼赞。于肃穆氛围中，拥足量的稳重质地，大气而轩昂，沁入震撼。双开电动门上，雕琢有精细考究，象征尊贵的中国帝制纹理，对门设计中介有金色"钱"字样将双半圆合为一圆，屋主姓氏即于此标注，如家徽般象征。

入殿究其视觉动线，如步过层层回旋过道，穿越无数楼阁屏栅，迎来的是气势中带着海纳百川的度量。再走进双通的回向廊道，看透光隔屏间投射出的眩目如挥剑，透过半圆弧隔栅导引光质，其影如展扇开来。光影展演随日照时程变幻强弱，映照在纹理清晰的灰石砖上。如此透过设计师的脉络铺陈，展现出空间里自然力与材质间的交融连结。

ARTTA
CONCEPT STUDIO

ARTTA 概念工作室

现代设计奖·优秀奖

ARTTA Concept Studio is an International multi-award winning innovative interior architecture design firm based in Hong Kong, established in 2010 by Arthur Tang. He graduated at The University of New South Wales with a Bachelor Degree in Interior Architecture.

The philosophy behind ARTTA comes from the word ART that defines our soul and with Architecture being our intelligence, combining these together bringing ART To Architecture.

Our team are built from talented individuals who have been trained globally and have unified together to work seamlessly to create projects with craftsmanship, timelessness and endless possibilities; designs that have their own characteristics.

　　ARTTA 概念工作室是一家国际知名的创新型室内设计公司，屡获殊荣，总部位于香港，于 2010 年由 Arthur Tang（亚瑟·唐）创立。他毕业于新南威尔士大学，获得室内建筑学士学位。

　　ARTTA 背后的理念来自于"艺术"这个词。艺术定义了我们的灵魂，建筑是我们的智慧，将二者结合，就是将艺术带入建筑。

　　我们的团队由来自世界各地的专业而富有才华的设计师齐聚而成。他们亲密无间地协作，创造出工艺精湛、永恒和拥有无限可能的项目。每一个设计都有自己的特色。

StagE
StagE 影院

设计机构：ARTTA 概念工作室
设计师：Natalie Chan, Arthur Tang, Louisa Tong, Cherry Ho, Ferlycia Man
客户：嘉禾
地点：中国香港
面积：1,400 m²
摄影：Gigantic Forehead

StagE is a new brand of Golden Harvest Cinema that is situated in Tuen Mun Town Plaza. Our concept is to combine 20th Century French Art Deco with a Contemporary style in the cinema design, giving a brand new experience to the guests.

The concept of "Chandelier"

A huge deformed Chandelier as the main feature of the box office, the light cast reflections on the floor and ceiling to diffuse artistic atmosphere throughout the interior. The magnificence and elegance become the attraction and that's where the focus point stands.

VIP room — "BackstagE"

At the back of the box office, customers may sneak peek the VIP room called "BackstagE", where craft glass and crude wood interior are used as insulation layer, creating a looming effect which triggers customer's curiosity and ambitions.

Theatrical design of cinema houses

To provide an exclusive experience, we applied theatrical design concept within the cinema houses. This creates a harmonized and solemn atmosphere to allow audience to indulge in the movie not just as an entertainment but also an artistic activity.

StagE 是嘉禾电影的新品牌，进驻于屯门城市广场。我们的理念是将 20 世纪的法国装饰艺术与当代风格相结合，在电影院设计中为客人带来全新的体验。

"枝型吊灯"的概念

一个巨大的变形枝形灯是售票区的主要特色。它的光线投射在地板和天花板上，让整个室内弥漫着艺术氛围。华丽和优雅是它的迷人所在，也是焦点所在。

贵宾室——"BackstagE"

在售票区后方，顾客能够窥见被命名为"BackstagE"的贵宾室。它采用工艺玻璃和原木作为隔层，产生了一种若影若现的效果，诱发着顾客的好奇心和欲望。

电影院的剧场设计

为了提供独特的体验，我们将剧场设计理念应用于电影院，创造了一种和谐而庄严的气氛，让观众的观影不仅是一种娱乐，也是一种艺术活动。

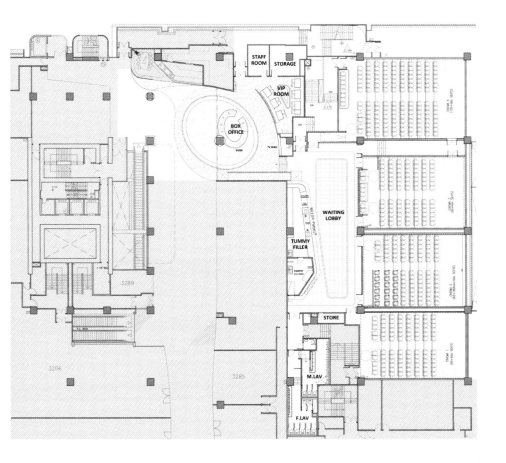

Cozi Lounge

悦品休闲厅

设计机构：ARTTA 概念工作室
设计师：Natalie Chan, Arthur Tang, Cherry Ho, Vince Wong
客户：悦品酒店
地点：中国香港
面积：188 m²
摄影：Gigantic Forehead

Situated inside Hotel Cozi Lounge will take you into another dimension. As the customers walk in from the grand and nature-themed reception with a day time ambience through into a much darker and evening-like environment restaurant and bar for people to relax and unwind in.

A unique pattern was created and used around the interiors surrounding, the using of the different metals to form layers and help enhance the height of this restaurant. We have decorated the restaurant and bar with an Art Deco theme in mind, with a combination of gold and different variety of stones for a sophisticated and elegant finish.

There is a design challenge that Cozi Lounge is located in an industrial area. To provide a relaxing atmosphere and environment to customers, several patterns and combination of materials are adopted to create a grand and elegant space, so that customers can enjoy their time in the lounge even in an urban area.

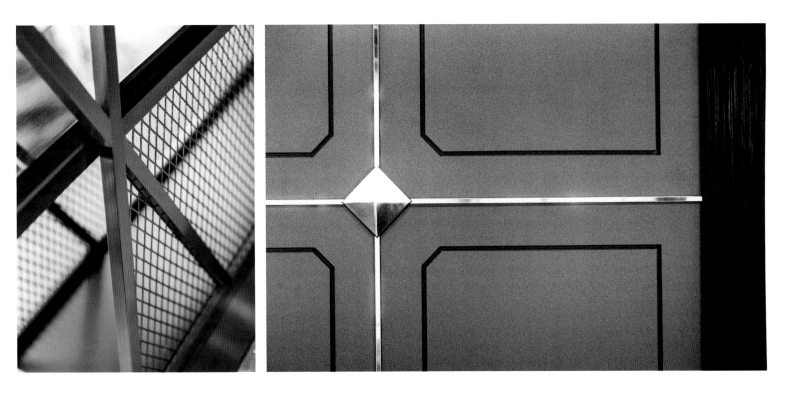

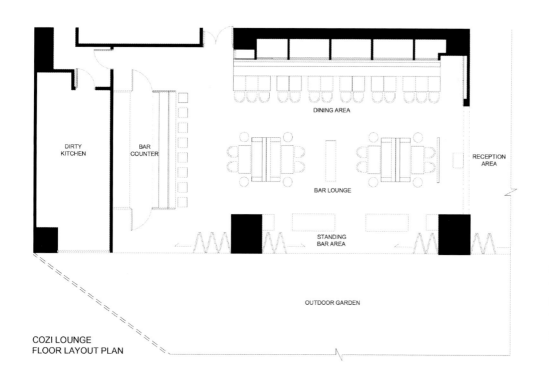

DINING AREA

DIRTY KITCHEN

BAR COUNTER

RECEPTION AREA

BAR LOUNGE

STANDING BAR AREA

OUTDOOR GARDEN

COZI LOUNGE
FLOOR LAYOUT PLAN

在悦品酒店内，悦品休闲厅带您进入另一个维度。顾客从如白天般恢弘敞亮的、以自然为主题的接待大厅走进这个更为黑暗、如夜晚环境的餐厅和酒吧，放松自我。

室内环境中使用了一种特别创作的独特图案，它通过不同的金属叠加来形成层次，帮助提高餐厅的视觉高度。我们以装饰艺术为主题来装饰餐厅和酒吧，将金色和不同种类的石头相结合，作为精致优雅的细节处理。

由于悦品休闲厅位于一个工业区，这给设计带来了挑战。为了给顾客提供一个放松的氛围和环境，我们采用了多种图案和材料的组合，创造出一个宽敞而优雅的空间，让顾客即使身处城市也能在这里拥有美好时光。

Silver Award ·
Overseas Winners of
Modern Design Award

海外区域
现代设计奖 · 银奖

ZOOI
Interior Studio

ZOOI
室内设计工作室

海外区域
现代设计奖 · 银奖

We create stylish and comfortable interior design.Any project of ZOOI studio is made individually. The main idea and concept for us is what will be right for a particular person, family, business. We rely on our experience, creativity and latest achievements to get the best solution for your interior. The result is always better than what the customer could have imagined.

Design affects human. It changes the view of life and the self-image.

According to the opinion of ZOOI specialists, a good design is a space that matches with the lifestyle, habits and nature of the person. It gives the feeling that the environment has been chosen especially for that person. The stylish and organized interior represents the way to feel happy and complete. Design and comfort are primarily important for the success of a person's life.

Every project of our design studio is developed in collaboration with the client. All design elements, designed by the experts of ZOOI, correspond with the wishes of the customer. Whatever it is, we fulfill the task professionally with attention to details. The client can be sure: a beautiful stylish object created for him by ZOOI designers is always a high quality and practical decision.

　　我们创造时尚舒适的室内设计。ZOOI 室内设计工作室的任何项目都是专项执行。我们的主要设计想法和概念是探寻对一个特定的人、家庭或企业，什么才是合适的。我们依靠我们的经验、创造力和最新的成就，为客户制定最好的室内设计解决方案。我们所呈现的结果总是超出客户预想。

　　设计影响人类。它能改变人们的人生观和自我形象。

　　ZOOI 的专家认为，好的设计应该是与使用者的生活方式、习惯和气质相匹配。它给人的感觉是这个空间环境是专门为他搭建的。时尚且精心布置的室内设计代表着人们感受快乐和圆满的一种方式。设计感和舒适性对人生的成功至关重要。

　　我们设计工作室的每一个项目都是与客户合作开发的。所有设计元素均由 ZOOI 专家设计，以完美符合客户意愿。无论是什么委托，我们都以专业的态度和对细节的高度重视来对待。客户可以确信的是：ZOOI 设计师所提供的精美设计永远是高品质与实用性的结合。

River Stone

雨花石

设计机构：ZOOI 室内设计工作室
设计师：Pavel Voitov, Maxim Doschinsky
地点：Kiev
摄影：Andrey Bezuglov

To implement this project, three apartments were merged into the River Stone by ZOOI interior studio. As a result, this apartment, conceived for a young family, occupies a half-floor of 300 square meters.

The open space, which includes a living room, kitchen and dining room, is conventionally divided by pylons, technical track lights from the XAL factory, colors and textures.

The kitchen, made in the shape of an island, allows you to use it for completely different purposes and consists of various storage systems, accommodating everything you need.

In the living room there is a huge modular sofa from the Italian factory Lema, with its coffee and milk colors. A coffee table of the Gallotti & Radice factory, and a floor lamp from the Italian factory Vibia.

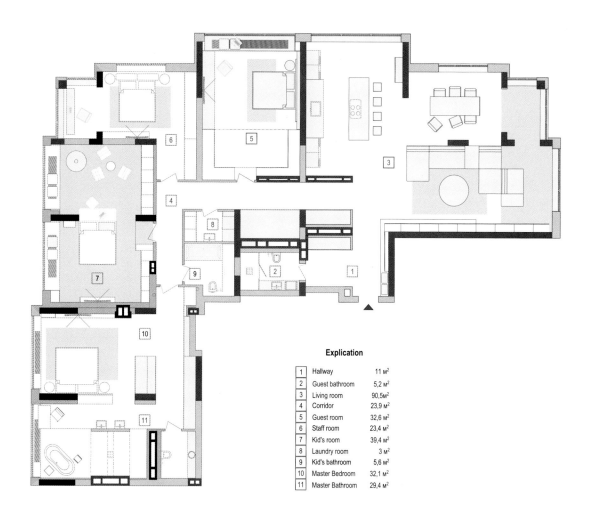

Explication

1	Hallway	11 м²
2	Guest bathroom	5,2 м²
3	Living room	90,5 м²
4	Corridor	23,9 м²
5	Guest room	32,6 м²
6	Staff room	23,4 м²
7	Kid's room	39,4 м²
8	Laundry room	3 м²
9	Kid's bathroom	5,6 м²
10	Master Bedroom	32,1 м²
11	Master Bathroom	29,4 м²

为实现这个项目，ZOOI 将三套公寓合并，成为如今这套河石公寓。这套为年轻家庭所设计的半错层结构公寓占据了 300 m²。

客厅、厨房和餐厅形成的开放空间按照传统的方式，由塔架、XAL 工厂的技术轨道灯、不同的颜色和纹理来进行分隔。

厨房的形状像一个岛，可以实现不同的功能。它配备有各种储物系统，可来满足你所有的收纳需求。客厅里有一个巨大的、出自意大利 Lema 工厂的米色模块化沙发。客厅还配有 Gallotti & Radice 工厂出品的咖啡桌和意大利 Vibia 工厂出品的落地灯。

Bronze Award ·
Overseas Winners of
Modern Design Award

海外区域
现代设计奖 · 铜奖

Dinara
Yusupova

海外区域
现代设计奖·铜奖

When dreams come true. Many years ago it was just a dream, nowadays the U—STYLE studio is a coherent, congruous team, where specialists such as architects, designers and engineers work in tune.

　　曾经遥远的梦想如今已经成为现实。许多年前，它只是一个梦想，但今天，U-STYLE 工作室已经具备一支密切协作的团队。在这里，建筑师、设计师和工程师等专业人士协同而作。

UI014

UI014 公寓

设计机构：U-Style
主案设计师：Dinara Yusupova
家具设计：Gamma Dandy, Molteni & C, Gamma
灯光设计：Delta Light, Deligtfull, Vibia, Fambuena, Oluce, Artemide
门和玻璃隔断：Rimadesio
面积：164 m²
摄影：Sergey Polyushko

Maximum attention has been paid to the comfort and the quality. Colors and textures has been delicately chosen. Ceramic panels and plaster have been used.

Thanks to the light play the project turned out bright and unforgettable.

项目品质和舒适性被赋予了最大的关注。在本案项目的设计中，颜色和纹理都经过精心挑选。材料上使用了陶瓷面板和石膏。

加上灯光的作用，这个项目变得明亮而难忘。

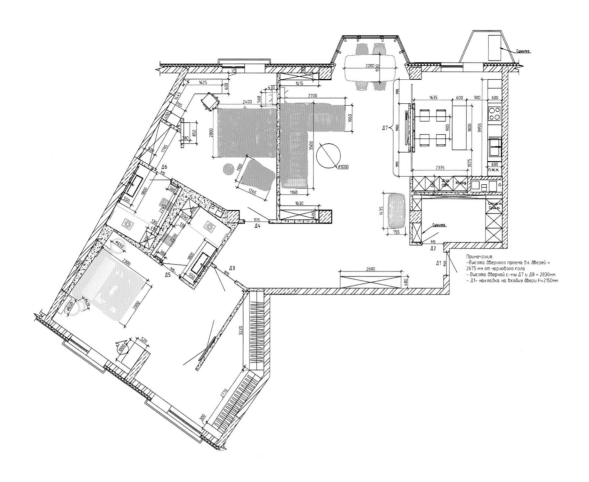

Примечания:
- Высота дверного проема вн. дверей =
  2675 мм от чернового пола
- Высота дверной с-мы Д7 и Д8 = 2830мм
- Д1- накладка на входные двери Н=2150мм

UI009

UI009 公寓

设计机构：U-Style
主案设计师：Dinara Yusupova
地点：乌克兰基辅
面积：100,42 m²
家具：Poliform, Minotti, Viccarbe, B&B Italia
灯光设计：Inarchi, Slv, Flos, Hold pendant lamp SKLO
水暖：Laufen, Villeroy&Boch, Hidra
摄影：Andrey Bezuglov

100 m² of room for thought, ideas and thorough process. A comfortable working place, an appropriate rest zone and leisure activities areas with all the amenities of living in a modern functional interior have been created inch by inch. Thus the combination of texture, wood and concrete create the style, congruously and unimposingly turning the flat into a modern place you will find it pleasant to spend time in. The coziness, lack of unpleasant imposing distractions, annoying variety and plentifulness of colour got combined in minimalistic interior design "UI009".

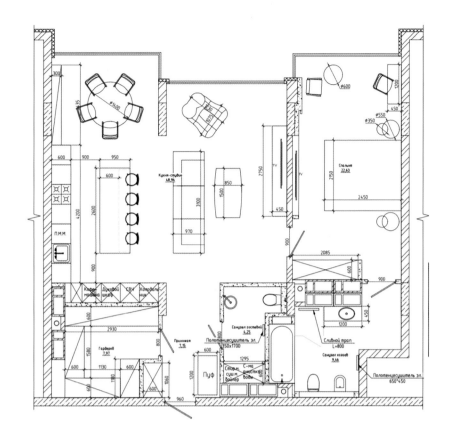

本案设计对 100 m² 的空间进行思考、赋予创意和作全面处理。一个舒适的工作场所，一个恰到好处的休息和休闲活动区，配备上居住在现代化、功能齐全的室内空间所需的所有生活设施，被一寸一寸创造了出来。纹理、木材和混凝土的结合形成了自我风格，和谐而又完美地将原公寓变成一个现代居所，令人在这里度过的时光总是那么愉悦。舒适且没有令人不快的干扰和令人烦躁的复杂色彩成就了这个简约的室内设计项目"UI009"。

AVIRAM–
KUSHMIRSKI

AVIRAM-
KUSHMIRSKI
设计工作室

海外区域
现代设计奖·铜奖

Our studio deals with all aspects of interior design: planning, design and decoration of homes, offices, shops and commercial spaces. We aim to design spaces that are designed in a uniform and consolidated matter, whilst reaching the highest aesthetic standards possible. We strive to create spaces that have their own personality and at the same time are distinctly unique in their own design, whilst taking into consideration the client's needs and the constraints of each particular space. At the same time we take into deep consideration the personality of our clients and we create a space that is not just functional but that also reflects the lifestyle and the character of our clients, as individuals. Through our careful attention to details we are able to create a space. With its own design philosophy, distinctly unique to each and every project, our service is comprised of assessing our client's needs, making a detailed work-plan and timetable for both design, decoration and home styling , until the very last element is in place. At the same time, we oversee the entire project from in inception until its completion. We are available to our clients at any time after the conclusion of the project, as needed, for any matter however large or small. Not only do we love design, but we enjoy creating spaces that will stand the test of timeless classic design that is not only accurate, but that will be relevant for many years to come as well as now.

　　我们的工作室涉及室内设计的各个方面：住宅、办公室、商店及商业空间规划、设计和装饰。我们力求空间设计协调统一，同时达到最高的审美标准。在考虑客户需求和每个特定项目的局限性的同时，我们努力创造拥有自我个性、且在设计上拥有独到之处的室内空间。同时，我们深入考虑客户的个性特征，让我们所创造的空间不仅是功能性的体现，也能反映客户作为一个独立个体的生活方式和个性。对细节的细心关注让我们有能力创造空间。秉承着自身的设计理念，我们的每个设计与项目都有其独特之处。我们的服务包括评估客户的需求，为设计、装修和家居设计制定详细的工作计划和时间表，直到最后环节。同时，从项目开始直到完成，我们监督并会跟进整个过程。即使在项目结束后，无论何时，若有需要，无论事件大小，我们都可以为客户提供后续服务。我们不仅热爱设计，也渴望创造出经久不衰的经典空间——它们不仅规范标准，而且对现在乃至未来许多年后的设计都具备参考性。

Apartment in an Older Building in the Heart of Tel Aviv

特拉维夫市中心旧房公寓改造

设计机构： Aviram – Kushmirski 设计工作室
设计师： Oshri aviram & Dana Kushmirski
客户： 一对年轻的父母
地点： 以色列特拉维夫
面积： 92 m²
摄影： Oded Smadar

The concept we chose for the design of this apartment – limited edition.

In practice, we were inspired by the clients' list of requirements. They were looking for designers providing personal, customized design, just for them and expressing their individualistic life style.

Our commitment to creativity, design and love for people spurs us towards original, creative thinking that challenges us in every new project.

Throughout the process, we seek the different, without falling into the convenient trap that is the obvious and the safe.

The acute need for original thinking creates the unique difference instilled into every project.

We use only minimalist, natural materials in every project. Those natural materials are worked by specialist artisans and their richness forms a precise, composite whole. We ensure that all the strengths of those materials are expressed.

The three main materials form a cascade of metals, wood and stone combined as a range of materials, displayed as surfaces and as significant masses with different finishes. Together they create a work of art comprising layers and blocks, small details and broad statements; a spectrum of textures. Each natural material has its own language in perfect synergy with the distinct language of every other material. In the master bathrooms we used black and white marble in geometric tiles that were customized for this apartment. In the guest bathroom we used other shape of geometric tiles with a combination of dark grey marble.

The apartment was paved with natural parquet in "fishbone" shape.

When selecting the processes through which natural materials are implemented, we feel that it is vital to make minimal use of "readymade" and we accentuate manufacturing processes employing precious artisanal inputs displaying the artistic value in the traditional skills found locally.

The results express our enthusiasm and respect for natural materials and our never ending efforts in the examination of boundaries and to go beyond the boundaries of natural materials. All that comes together in our design approach, which is based on "honest design", emphasizing the purity and practicality in natural materials. This is organic design inserting nature itself into the space to create a feeling and a look that resonates with vitality and freshness.

我们为这个公寓设计所选定的概念为"限量版"。在实践中，我们受到客户需求清单的启发。他们正在寻找能够提供个人的、定制设计的设计师，帮助他们来表达他们充满个性的生活方式。

对创作和设计的使命以及对生活的热爱激励着我们探寻原创和创新思维，并在每一个新项目中挑战自我。

一直以来，我们总在寻求"不同"，让自己不落入那些讨巧便捷却俗套的陷阱中，即使那些方式唾手可得且更为安全。对原创思维的迫切需要让我们将

独创性注入了每个项目。

我们在每个项目中只使用极简的天然材料。这些天然的，形态复杂多样的材料经由专业工匠制作，成为一个精致的复合整体。我们要确保这些材料的所有优点都能得到充分使用效果。

三种主材的使用，形成了金属、木材和石头的层叠，它们的结合作为一种材料组合，不仅充当着饰面，也体现了一个空间所拥有的不同细节处理。它们共同形成了一件艺术作品，有层次、有分区、有细节、有格局，充满纹理感。

Our designs are minimalist, with clean lines and forms based on precision and careful tuning of the fine details. We will always aim to create a design that is timeless, creative, fresh and modern. Very importantly, our designs are highly germane to the convenience modern lifestyles can provide.

每种自然材料都有自己的语言，并与其他材料的独特语言完美契合。在主卧浴室，我们使用了为这套公寓定制的黑白大理石几何瓷砖。在客卧浴室，我们使用了其他形状的几何瓷砖，并搭配使用深灰色大理石。公寓铺设了"鱼骨"形状的天然拼花地板。

在考虑如何运用这些天然材料的过程中，我们认为尽量减少"现成品"的使用至关重要。我们十分看重生产过程中能够加入珍贵的手工艺元素，以此显示当地传统技术的艺术价值。

项目成果表达了我们对天然
材料的热情和尊重，以及我们对材
料边界性的不断探索以及渴望超
越这些边界的不懈努力。所有这些
都体现在我们的设计理念中——
"诚实设计"，关注自然材料的纯
粹性和实用性。这是一个有机的
设计，将自然本身融入到空间中，
它所呈现的感觉和外观能够与自
然的活力和新鲜产生共鸣。

我们的设计是极简的，基于
精确和细致的细节调整，有着简明
的线条和形式。我们将始终致力于
永恒、创新、清新和充满现代感的
设计。还有一点非常重要，那就是
我们的设计要能充分贴合现代生
活方式的追求便捷性的本质。

Excellence Award · Overseas Winners of Modern Design Award

海外区域
现代设计奖·优秀奖

33BY Architecture is the architectural workshop headed by the architect - Ivan Yunakov.

Workshop 33BY Architecture strongly since 2007 was fixed in leaders of the architectural market of Ukraine, advancing the image and an esthetics of comfortable and functional housing.

Distinctive feature of activity of the company – a full cycle of design, from preliminary expert examination of the site under construction, the subsequent development of all design stages, coordination, developments of interiors and conducting designer's service.

Having the strategic development plan, and using in work of technology of project management the company has accurate structure business – processes that provides the required term and quality of implementation of the project.

The company involves in the work on outsourcing only the highly professional licensed companies, such as MEP Engineering.

Architects and designers of a workshop, constantly study and master experience of domestic and foreign architectural bureaus and workshops, increase the professional level and the last achievements in the projects developed by the company implement.

Studying our objects it is possible to pay attention that the field of activity of the company includes design of housing, office, trade, hotel, administrative constructions, cottages not only in Kiev, but also in other cities of Ukraine, Europe and CIS.

33BY Architecture

33BY 建筑事务所

海外区域
现代设计奖·优秀奖

33BY 建筑事务所是由建筑师 Ivan Yunakov 带领的建筑工作室。

33BY 建筑事务所自 2007 年以来稳稳占据着乌克兰建筑设计市场的引领者地位，不断提升着舒适及功能性住宅的形象和美学。

公司事务的特色之一在于，从前期专家对施工现场的考察，到后期各设计阶段的开发、协调、设计，再到设计师的服务，提供了一整个完整的设计周期服务。

公司具有战略发展规划，并在工作中运用项目管理技术。它具有明确的业务流程结构，为项目验收标准和所需期限提供了标准。公司所参与的项目仅外包给高度专业化的认证公司，如 MEP 工程。

工作室的建筑师和设计师们通过不断学习和掌握国内外建筑事务所和工作室的经验，来提高专业水平，并在公司开发的项目中贯彻落实。

通过对我们项目的研究，你能发现公司的工作领域不仅包括在基辅的住宅、办公、贸易、酒店、行政建筑、行政区域、别墅的设计，而且还包括在乌克兰、欧洲和独联体的其他城市的设计项目。

Buddha House
佛·住宅

设计机构：33BY 建筑事务所
设计师：Ivan Yunakov, Olga Kornienko, Katrich Jaroslav, Timothy Bogatyrenko
瓷砖：Stone (onyx) ceramic tiles, grey ceramic tile, stone texture ceramic tile, wood texture ceramic tile
地点：乌克兰基辅
面积：170 m²
摄影：Oleg Stelmakh

Interior of a small summer house for a very interesting, young and happy family that follows Buddhism, enjoys yoga and loves to travel a lot. The building is intended for leisure and entertaining.

The building is decided in wood, glass and metal. Just architectural composition is maintained outdoor pergolas and terraces. The interior used natural wood, stone, copper, textiles and varied game world, both natural and artificial.

这个小避暑别墅是为一个非常有趣、年轻、快乐的家庭而做的设计。他们信奉佛教，喜欢瑜伽，喜欢旅游。这个空间被用于家庭成员们的休闲娱乐。

这座建筑物由木头、玻璃和金属建造而成。建筑室外的凉棚和露台也是建筑的组成部分。室内采用天然木材、石材、铜和纺织品装饰，还有各类自然形成或人工搭建的游戏空间。

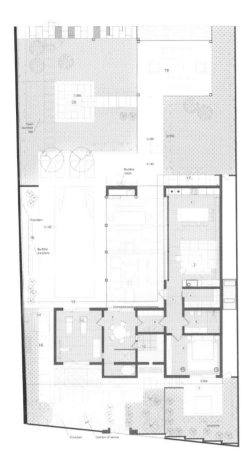

Explication

#	Name of premise	S/sq m
1	Corridor	6.3
2	Living room	21.0
3	Kitchen-dining	23.5
4	Wardrobe	5.8
5	Bathroom	7.0
6	Bedroom	18.1
7	Terrace with Jacuzzi	10.5
8	Dressroom	5.25
9	Dayroom	14.1
10	Sauna	6.1
11	Gym	20.2
12	Main terrace	56.9
13	Terrace with swimming pool	65.5
14	Summer open shower	2.9
15	Relaxing zone	7.9
16	Fountain	10.5
17	Relaxing zone	5.2
18	Relaxing zone	9.6
19	Terrace of main building	34.7
20	Terrace with open fire	
	Total area of premises	127.4

*Atelier Design
N Domain*

Atelier Design
N Domain 工作室

海外区域
现代设计奖 · 优秀奖

A Kaleidoscope creates a certain amount of dynamism which challenges the static thus permitting a dialogue of possibilities. At Atelier Design N Domain, co-founded by Ar. Anand Menon and Ar. Shobhan Kothari, the partnership has encouraged dialogue on design possibilities and challenges that come with it thus creating visually striking spaces delivered within the parameters of pragmatism. The designers engaged at the firm further propel this idea of continuance and prevent stagnation of thought.

Atelier Design N Domain is an architecture + interior firm doing boutique work in the genre of residential, corporate, hospitality and retail. The firm has successfully completed 16 years in practice and has a growing body of work with some of the renowned names of the society. The firm has a team of 25+ designers aligned and determined to create a niche in the field of architecture and interior and to meet with client satisfaction. At Atelier Design N Domain, design is treated as a process; something arrived with due consideration of pragmatism but not limited by them. The firm dwells in atmosphere of "studio" practice combined with a sense of realism.

The firm constantly strives for a clarity of thought when it comes to the realization of spaces be it architecture or interiors. Our belief that we are as good as our last completed project has persistently pushed us to strive for better. Our self-critical nature helps us to take compliments and criticism in the same stride.

After completion of over a decade in practice, the firm believes it retains "heart" of a child but with vast experiences to guide its future clients/projects.

　　万花筒能够创造动态感，挑战静态，两者之间产生的交流碰撞赋予设计更多的可能性。Atelier Design N Domain 是由 Ar. Anand Menon 和 Ar. Shobhan Kothari 共同创立的工作室。二人就设计的可能性和随之而来的挑战进行对话，从而在实用主义的参数下创造耳目一新的空间。工作室的设计师们又进一步开阔思维，推动了这种理念的延续。

　　Atelier Design N Domain 是一家从事住宅、企业、酒店和零售领域精品业务的建筑＋室内设计公司。该公司已成功地走过了 16 个年头，并与社会上的一些知名品牌建立起越来越多的工作往来。公司团队拥有超过 25 名设计师，他们有着共同的决心，力求在建筑和室内设计领域开拓自己的市场，做到令客户满意。在 Atelier Design N Domain，设计被视为一个过程。这个过程将实用主义考虑在内，却并不受它们的限制。该公司充满一种"工作室"结合现实主义的氛围。

　　当涉及到空间的实现时，无论是建筑还是室内项目，公司都在不断地追求清晰的思路。我们相信我们的每个项目都会完成得同前一个项目一样好，而这一信念能推动我们一直努力做得更好。我们自我反思的品质帮助我们能够同时接受赞扬和批评。

　　经过十多年的实践，他们相信自己依然保有"初心"，而丰富的经验能更好地指引他们完成未来的客户项目。

OHANA Residence
OHANA 住宅

设计机构：Atelier Design N Domain
地点：印度 古吉拉特邦 瓦多达拉
面积：1,281 m²
摄影：©Photographix | Sebastian + Ira

"OHANA" a Spanish word, meaning "family bond", is an attempt by us to design a house with spaces and experiences that resonate the essence of spending time together as a family. Set in an environment of farmland the intent was to come up with a design that establishes a strong inside outside connect. A house that attempts to maximize the use of natural light for the interior spaces and address the various aspects of a contemporary modern lifestyle.

The entrance driveway and vehicular drop off is characterized by calibrated granite clad walls and a fluted metal screen towards the garden side. The otherwise blank entrance facade with a singular door marks the entry point to the house. Strategically located in alignment with the entrance door is a larger than life brass sculpture by Navin Chahande. Taking into consideration the clients brief and various site forces, the plan is essentially a stretched linear box aligned with the cardinals to ensure that all functions receive ample morning sunlight and enjoy undisturbed views of the adjoining landscape. At the core of its planning is the elongated verandah space that not only receives you when you enter the house but also acts as the connect between all functions at the ground level. The entrance door is flanked by a skylit visitor waiting area on the right and mandir to the right. One has to traverse a skylit waterbody to access the mandir. As you move along, you are received by a stunning ceramic and metal art installation by Gopinath Subbanna. The artwork not only creates a sense of pause but also announces the entry

point to the formal living on the opposite side. The formal living is a taller rectangular volume with floor to ceiling corner glazing that penetrates into the landscape beyond. The luxurious verandah space is characterized by the informal seating juxtaposed against the bar finished in micro concrete and high sheen corrugated copper. Soft mood lighting translates the naturally lit verandah by day into a high energy party zone at night with a seamless connect to the outdoors. Located off the verandah is the all black slate clad powder bathroom that adds to the lounge like experience.

Large sliding doors separate the verandah space from the informal family living space. The Centrally located family living acts as a connect between the master bedroom suite and guest bedroom on the ground floor and the staircase that leads to the upper floor.

The kids bedrooms on the upper floor are punctuated by the media room which also doubles up as their lounge space. Dark tones in the furniture, fabric clad walls and ceiling punctuated by track lights add to the much desired theatrical experience. The extension of the media room is a all wood clad balcony. The media room announces itself on the exterior as a cantilevered wood clad box sandwiched between the floor slabs. It floats above the swimming pool like a diving board, in perfect alignment, inviting one to take a plunge.

The interiors of the kids bedroom couldn't have been more like chalk and cheese. the son's bedroom in hues of grey, dark wood tones and camel tan bed back clearly project a masculine taste and aesthetic. In sharp contrast to this the daughter's bedroom explores a muted aesthetic with white stained oak wood for furniture and a very English setting. Her choice of fabrics and embellishments clearly make for a distinct style statement.

The landscape design for the house attempts to compliment its open plan. The master plan for the landscape is broadly divided into the formal party lawn directly accessible from the drop off driveway and the family lawn that runs parallel to the house. Minimalism in nature the design features a large pool with a built in pool bar housed between strategically located frangipani trees. The adjoining wooden pool deck boasts of the all glass gym on one side and the shaded gazebo on the other. Aligned with the pool bar, the gazebo acts as an informal hang out space to soak in the surrounding greens. The Fluted solid wood backdrop of the gazebo features a concealed door that leads to the in house spa and changing facilities. The landscape also features a 40 feet floating granite bench element with a backdrop of staggered calibrated granite walls and pampus grass.

　　"OHANA"是一个西班牙语词汇，意为"家庭纽带"。这一项目中，我们试图设计一所住宅，令它的空间和体验能与家人共度时光的精神产生共鸣。在一个田园环境中，我们想要通过设计实现紧密的内外连接。住宅的室内空间试图最大限度地利用自然光，并能满足现代生活方式的方方面面。

　　入口车道和车辆落客点以统一的花岗岩覆盖墙和面向花园一侧的凹槽金属屏幕为特征。除此之外，带有一扇单门的空白入口立面标志着房子的入口点。Navin Chahande 所设计的比真人还大的黄铜雕塑巧妙地与入口门对齐。考虑到客户的需求和不同的场地约束，该设计从根本上来说是一个与基点对齐的延

展的线框，以确保所有功能区都能获得充足的早晨阳光，并享受毗邻的自然景观。其规划的核心是狭长的阳台空间，使它不仅能够在你进入住宅时迎接你，而且能够充当一层各个功能区之间的连接。入口门的右侧是一个开有天窗的访客等候区和一个朝右的敬神室。访客需要穿过由天窗照亮的水域才能进到敬神室。当你继续前进时，迎接你的是 Gopinath Subbanna（高佩）令人惊叹的陶瓷和金属艺术装置。这个艺术作品不仅给人一种赞叹之感，也是对进入对面正式生活区域的一种宣告。

正式的生活空间是一个更高的矩形体量，落地玻璃的使用让这一空间能够渗透进前方的景观中。豪华的阳台空间中配备有休闲座椅，放置在由微粒混凝土和高光泽波纹铜制成的吧台边。夜晚，柔和的氛围灯将白日采光充足的阳台

转变成充满活力的派对场所，与户外空间无缝相连。位于阳台外的是全黑石板覆盖的盥洗室，增添了一种贵宾厅的体验。

大型推拉门将阳台空间与日常家庭生活空间分隔开来。一楼位于中心位置的家庭生活区连接起主卧室套房和客房，以及通往上层的楼梯。

楼上的儿童卧室之间安置有媒体室，媒体室也兼作他们的休息室。深色调的家具，织物包覆的墙壁和天花板上点缀着轨道灯，为空间增加了非常契合的戏剧体验。媒体室的延伸部分是一个全木质的阳台。媒体室从外部看是一个悬挑着的木质包层盒子，夹在楼板之间。它像一块跳水板一样漂浮在游泳池上方，呈完美的直线型，好像在吸引着人们跳入水中。

孩子们卧室的内饰风格迥异。儿子的卧室使用灰色调，深色木材和驼色的床背，清楚地投射出阳刚气和美感。与此形成鲜明对比，女儿的卧室则探索了一种柔和的美学，使用了白色的橡木家具和非常英式的配置。面料和装饰的选择呈现出一种独特的风格。

住宅的景观设计尝试对建筑的开放空间致意。景观的总体规划大致分为正式的派对草坪和家庭草坪两部分，前者可以直接从落客车道进入，后者与房子紧邻。设计以极简自然主义为核心，亮点在于一个带有酒吧的大型泳池。这个酒吧被安置在鸡蛋花树之间，而鸡蛋花树的分布也经过了精心设计。毗邻的木制泳池平台的一侧是全玻璃健身房，另一侧是遮阳露台。露台与泳池酒吧对齐，作为一个轻松的休闲空间，让人们能够沉浸在周围的绿意中。露台的凹槽实木背景以一扇隐蔽的门为特色，这扇门通向室内水疗中心和更衣室。这个景观区还配有一个 12 m 的悬空花岗岩长凳，背后是交错统一的花岗岩墙壁和蒲苇草地。

SHH Architecture
& Interior Design

SHH 建筑与室内
设计工作室

海外区域
现代设计奖 · 优秀奖

Founded in 1992 by David Spence, Graham Harris and Neil Hogan, SHH Architecture & Interior Design is an internationally acclaimed London-based practice that is dedicated to working in residential and commercial fields with integrity, intelligence and flair. Providing expertise in architecture and interiors ranging from Super Prime homes and luxurious residential development to hospitality environments in large arenas, the team comprises 50 talented individuals from around the globe, who bring rich diversity in thinking and creative design.

  SHH 建筑与室内设计事务所由 David Spence（大卫·史宾斯），Graham Harris（格雷厄姆·哈里斯）和 Neil Hogan（尼尔·霍根）创立于 1992 年，是一间位于伦敦的国际知名设计事务所。它以其诚信、智慧和才华，致力于住宅和商业领域的设计工作。从顶级住宅和豪华住宅开发到大型酒店环境设计，SHH 建筑设计事务所在建筑和室内设计方面提供专业服务。其团队由来自全球不同国家的 50 名人才组成，为公司带来了多维度的思考和创造性的设计。

West London House
西伦敦住宅

设计机构：SHH 建筑与室内设计事务所
客户：Gilliam Properties
地点：英国伦敦
面积：465 m²
摄影：Alastair Lever

For developer Gilliam Properties, SHH was commissioned to transform and add value to an existing, 5,000 sq ft, detached and freehold property at a prestigious address in the Holland Park area of West London. Planning permission was obtained to extend, remodel and fit out the property — formerly divided into flats — in order to create a single and luxurious 9,000 sq ft family home, retaining the building's facade, but demolishing the remainder of the existing structure. The result is a 5-storey, luxury home which offers its purchaser bright, spacious and well-proportioned accommodation, together with a 107-feet-long southwest-facing rear garden and one of West London's largest indoor pools.

开发商 Gilliam Properties(吉列姆地产公司) 委托 SHH 对其位于伦敦西部 Holland Park(荷兰公园) 的一处地产进行改造增值。该地产位于黄金地段，是一个有着 465 m² 的独立的、有着永久产权的地产。该项目获得了规划许可，允许对这处原被划分为多个公寓的房产进行扩建、改造和装修，将它变为一个面积 836 m² 的独立豪华家庭住宅。建筑的原本立面会被保留，但建筑结构的其余部分会被拆除。最终建成一栋 5 层的豪华住宅，能为住户提供明亮、宽敞、划分合理的居住空间，以及一个 1.83 m 长的、西南朝向的后花园，和一个西伦敦最大的室内游泳池（之一）。

Evaluation Criteria of
Modern Design Award
现代设计奖评审标准

一、参赛类别

展现现代风格的实景案例：住宅空间、商业空间、酒店工程、环境艺术空间

二、奖项评审标准

现代设计奖：

1、作品属于原创，聚焦现代风格，注重创新；

2、极具现代家居素色美学、结构美学，兼具个性、功能性、实用性；

3、综合设计性强，在造型结构、色彩运用、软装搭配、美学内涵等多维度上把握和谐，手法独到；

4、在现代简约风格设计上具有示范性意义。

现代设计·最佳创意奖：

1、作品属于原创，聚焦现代风格，注重创新；

2、设计理念独到，具有美学诉求和潮流指向意义；

3、以创新思维意识，进一步挖掘和激活设计材料不同组合方式；

4、在作品的创意和风格上，能充分地表现出设计师的个性。

现代设计·色彩表现奖：

1、作品属于原创，聚焦现代风格，注重创新；

2、以素色美学为基础，色彩搭配具有灵动感、和谐感；

3、具有丰富细腻的色彩表现手法，传达作品的设计情绪；

4、深刻理解"色彩是最直接的视觉语言"，通过创作意象在观赏者的视觉神经中留下长久印象。

现代设计·人文美学奖：

1、作品属于原创，聚焦现代风格，注重创新；

2、具有人文精神，对传统文化与现代生活方式进行传承与结合，并实现了有机融汇；

3、具有极强的艺术性和深厚的文化内涵，能体现设计师自身风格的独特美感；

4、以新旧融合为主轴，打造兼具文化情趣与设计美学的现代创意风格。

三、获奖设计师将加入国际青年设计师人才计划

1、参加大师执教的专业室内设计高级研修课程；

2、享有国际顶级专业设计交流访问机会；

3、获奖作品亮相国际青年设计师协会全球新锐设计师作品展；

4、获奖作品将收录至《素色美学空间：2018-2019 现代设计奖获奖作品集》，国际知名大师亲笔作序点评，全球六大洲 30 余国发行推广等。

⌐ARTPOWER⌐

致谢
感谢所有为本书做出重大贡献的设计公司和设计师，没有他们支持，本书将
不可能成功出版。同时感谢所有为本书默默付出的工作人员，他们也为本书
的出版提供了巨大的支持和帮助。

更多合作
我司长期致力于艺术、设计类图书出版，如果您有投稿意向，请将出版资料
发到：press@artpower.com.cn

首届"现代设计奖颁奖盛典"

点燃现代设计新生火苗

从 2018 年 4 月 10 日这个意义非凡的日子开始计算，
现代设计奖（2018-2019）从报名启动、作品征集、评选初审到复评入围，
覆盖 200 多个城市，从全国 3000 份优秀室内设计作品完成最后甄选。

盛典现场。

首届"现代设计奖颁奖盛典"隆重开幕，马岩松、黑川雅之等国际设计大师为现代设计奖得主荣耀加冕。